Textile Dyeing

Textile Dyeing

Editor

Abdulkerim Macar

scitus
academics

Textile Dyeing

Edited by **Abdulkerim Macar**

ISBN: 978-1-68117-538-6
Library of Congress Control Number: 2016942938

SCITUS Academics LLC,
www.scitusacademics.com
Box No. 4766, 616 Corporate Way,
Suite 2, Valley Cottage,
NY 10989

Printed in United States of America on Acid Free Paper

Preface

Dyeing is a method which imparts beauty to the textile by applying various colors and their shades on to a fabric. Dyeing can be done at any stage of the manufacturing of textile- fiber, yarn, fabric or a finished textile product including garments and apparels. The property of color fastness depends upon two factors- selection of proper dye according to the textile material to be dyed and selection of the method for dyeing the fiber, yarn or fabric. Dyes can be used on vegetable, animal or man-made fibers only if they have affinity to them. Textile dyes include acid dyes, used mainly for dyeing wool, silk and nylon and direct or substantive dyes, which have a strong affinity for cellulose fibers. Mordant dyes require the addition of chemical substances, such as salts to give them an affinity for the material being dyed. They are applied to cellulose fibers, wool or silk after such materials have been treated with metal salts. Sulfur dyes, used to dye cellulose, are inexpensive, but produce colors lacking brilliance. Azoic dyes are insoluble pigments formed within the fiber by padding, first with a soluble coupling compound and then with a diazotized base. Vat dyes, insoluble in water, are converted into soluble colorless compounds by means of alkaline sodium hydrosulfite. These colorless compounds are absorbed by the cellulose, which are subsequently oxidized to an insoluble pigment. Such dyes are colorfast. Disperse dyes are suspensions of finely divided insoluble, organic pigments used to dye such hydrophobic fibers as polyesters, nylon and cellulose acetates.

Textile Dyeing discusses in detail several emerging topics on textile dyeing. This book will serve as an excellent addition to the libraries of both the novice and expert.

Table of Contents

Chapter 1

A Smart Multifunctional Water-Oil Repellent Finishing and Dyeing of Cotton / Polyamide 6 Blended Fabrics

E. R. El-Zairy, W. R. El-Zairy

Faculty of Applied Arts, Printing, Dyeing and Finishing Dept. Helwan Univ., Cairo, Egypt

ABSTRACT

Polyamide/cellulosic blends offer valuable possibilities for combination desirable performance properties, i.e. self-cleaning, water-oil repellent, crease recovery angle, handle, etc. The main aim of this study is to search for a novel practical method of carrying out water-oil repellent finishing/reactive dyeing of cotton/polyamide-6 (50-50) in one stage using knittex® FEL crosslinking agent , OLEPHOBOL®CP-U , water/oil repellent agent, trichloroacetic acid as a catalyst as well as reactive dyes of Remazol type. The treated fabrics were padded and then squeezed to a wet pick-up of 70%. After being dried, the fabrics were thermofixed at different temperatures (100-180°C) for various durations (1-5). Good results of the overall fastness properties were obtained ranging between 4 to 4-5.

KEYWORDS

Polyamide/cotton, Dyeing, Finishing

1. INTRODUCTION

Demands for comfort and healthy lifestyles make it desirable to have textile products with multifuncitonality. Incorporated function properties to textiles such as durable press, self-cleaning, antimicrobial and thermochromic characteristics, water and oil-repellent, all have potential industrial applications. [1-6]

Blends of polyamide man-made fibres, with Cellulosic fibres offer valuable possibilities for combining desirable performance properties, developing attractive appearance, and novel design. Recently functional textiles have attracted much attention in recent years, e.g. UV-protection, antimicrobial, anti-radiation, water-and oil repellent, self-cleaning, etc. [7-12] Although the importance of these blends is understood, they have not maintained their reasonable market share. This can be attributed in part to difficulties in wet processing including dyeing to union shades the prevailing problem assiated with the union dyeing of polyamide-6 and cellulose blends from the same dyebath is the preferential uptake dye by polyamide-6. Cellulose/polyamide blends are usually dyed by the two-bath method: for example in a first step by the pad-steam process to dye the cotton component and then by the exhaust process to dye the polyamide component, or vice versa. However, this process is time-consuming and laborious when dyeing large yardages. Single bath method are available whereby two diffferent reactive dyes (one substantative to the polyamide and the other substantative to the cellulose-) can be used to dye this blended fabrics but the result is that the cellulosic component is dyed perfectly satisfactorily but the polyamide component is only superficially dyed, in fact merely stained, i.e. the dye has not diffused into the fibres under this method. [13, 14] Trichloroacetic acid decomposes at about 100°C to give a free radical, in the following manner:

$$CCl_3COO^- \, H^+ \longrightarrow {}^{\bullet}CCl_3 + CO_2 + H^+ + e$$

$$H^+ + {}^{\bullet}CCl_3 \longrightarrow CHCl_3$$

Since the chloroform and carbon dioxide produced are volatile, they are rapidly removed from the dye bath. The striking influence of the chlorine atom(s) on the dissociation of the proton is simply due to its inductive effect, since the chlorine atom in C-Cl attracts electrons in comparison with hydrogen atom in C-H, the effect is relyed in the following manner:

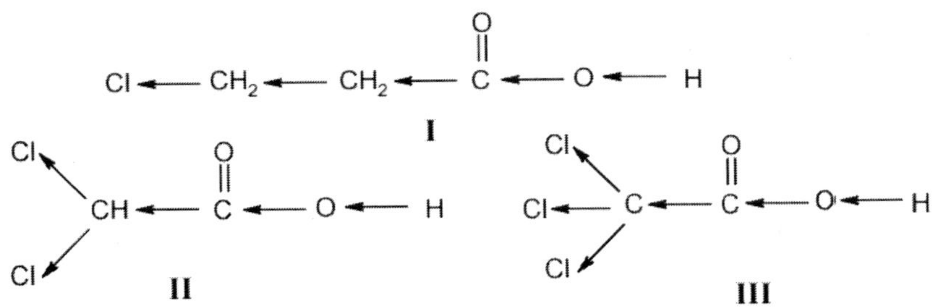

The decomposition of trichloroacetic acid, in either boiling dyebaths or during steaming and thermofixation giving a change in pH conditions from acidic to alkaline has been mentioned by Fabwerke Hoechst. [20-21]

In this work, we put forward a new method of carrying out water/oil repellent finishing and reactive dyeing of polyamide-6/cotton blend (Ny/Co) (50/50) in one step to evaluate the impact of the combined treatments on the functional and dyeing properties of the blended fabrics using the ability of trichloroacetic acid (TCAA) to change of the pH of the medium as the temperature changed, as well as to provide a process for dyeing cellulose/polyamide blends, which process makes it possible to dye both types of fibre continuously in a single bath and in one operation of dyeing and finishing.

2. EXPERIMENTAL

2.1. Materials

Mill-scoured, and semi-bleached cotton/polyamide-6 blended fabric (50/50 in yarn, 120 gm^2) was used.

Knittex® FEL cross linking agent (reactant crosslinking agent based on a modified dimethylol dihydroxyethylene urea, Huntsman), and oil, water and stain repellent agent, based on dispersion of a polmeric perfluorinated compound, nonionic/cationic, Huntoman, were of technical grade.

Commercial reactive dyes namely: Remazol Red X3BS, Remazol Blue RC, Remazol Golden Yellow RC as well as Remazol Blue RC, Dystar, were used.

Trichloroacetic acid (TCAA) and Na-carbonate were of laboratory reagent grade.

Hostapal® CV-ET (nonionic wetting agent based on alkylarylpolyglycolether, clariant) was of commercial grade.

2.2. Methods

2.2.1. Concurrent Functional of Ny/Co Fabrics

Bifunctional finishing and dyeing of polyamide-6/cotton (Ny/Co) blends were carried out as follows: the blended fabrics samples were padded twice through solution containing Remazol X3BS (20g/l), knittex® FEL crosslinking agent (0-80 g/l). OLEOPHOBOL® CP-U, oil, water and stain repellent agent (0-40 g/l) and trichloroacetic acid (TCAA) (0-4 g/l), the fabrics were then squeezed to a wet pick up of 70%. After being dried at 100°/3min., the fabrics (Ny/Co) were thermofixed at varying temperatures (100-180°C) and for different durations (1-5 min.). Finally, the fabric was washed with cold, hot water then soaped with a solution containing nonionic wetting agent (Hostapal CV-ET) and 2g/l Sodium Carbonate followed by washing with hot, cold water and finally dried.

2.2.2. Testing and Analysis

2.2.2.1. Colour Strength

The colour strength K/S, of the dyed/ finished fabric samples were measured at the wave length of maximum absorbance using an automatic filter spectrophotometer and calculated by the Kubelka – Munk equation [24]:

$$K/S = \frac{(1-R)^2}{2R}$$

Where K is the absorption coefficient, S is the scattering coefficient, and R is reflectance of dyed/finished fabric samples at wavelength of maximum absorption, the highest the K/S value, the greater the dye receptivity of the fabric.

2.2.2.2. Fastness Properties

Evaluation of washing, rubbing, perspiration and light fastness properties were performed according to AATCC test methods: (61-1972), (8-1972), (15-1973) and (16A-1972) respectively.

2.2.2.3. Crease Recovery Angle

Crease recovery tester, type: FF.07. metrimpex, Budapest was used for the determination of the crease recovery angle.

2.2.2.4. Tensile Strength and Elongation

Tensile strength and elongation (warp and weft) was measured on the tensile strength testing machine type FMGW 500 (Veb. Thuringer industriewerk Raunstein) at 25°C and 66% relative humidity.

2.2.2.5. Repellency Rating

Water/oil repellency was evaluated according to AATCC test method 22-1989.

3. RESULTS AND DISCUSSION

Since the primary objective of this work was to investigate the technical feasibility of carrying out dyeing and multi-functional finishing of polyamide-6/cotton fabric blend in a single-step process. The variable studied include trichloroacetic acid (TCAA) as a catalyst, finishing agent, repellent agent as well as thermo fixation time and temperature to discover the optimal condition for currying out bifunctional finishing and dyeing simultaneously of polyamide-6/cotton blended fabric using different coloures of remazol dyes, results obtained along with their appropriate discussions are given below.

3.1. Trichloroacetic Acid Concentration (TCAA)

It is established that reaction of finishing agents with polyamide-6/cotton blended fabrics as well as the reactive dye must be catalysed [25]. Thus the efficiency of simultaneous dyeing and finishing would be expected to rely on the nature and amount of the catalyst used. Table 1 shows the effect of trichloroacetic acid (TCAA) concentration on the extent of simulataneously functionalization and dyeing of polyamide-6/cotton blended fabrics. Obviously, within the range examined, it is clear that, increasing trichloroacetic acid (TCAA) concentration up to 2g/l brings about an improvement in K/S values, crease recovery angle (CRA), handle, water oil repellency rating and overall fastness properties. In addition, increasing the trichloroacetic acid concentration beyond 2g/l results in a slight dcreament in the aforementioned properties, which be discussed in terms of highest penetration of the dye onto polyamide-6/cotton blend along degree of temperature during drying and curing (100° - 160°C) is reached, the acid disappears from the system by free radical decomposition reaction. Further increase in trichloroacetic acid (TCAA), i.e. beyond 2g/l has practically a marginal decreasing effect on the K/S values, and other performance properties, which may attributed to the hydrolysis of reactive dye. Therefore 2g/l TCAA was chosen as the optimal concentration of TCAA in bifunctional finishing and dyeing of Ny/Co blends. Since attachment of the reactive dye to cotton component in the blend occurs through reaction of the dye either with N-methylol finishing agent, which is already chemically bonded to cotton in acidic medium or the ractive dye attachment to polyamide-6/cotton blends in alkaline medium (Equi.1, 2, 6, 7, 8). During the heat treatment of curing, decomposition via demethylolation of N-methylol compound occurs, depending on the catalyst concentration. As a result, amide groups are formed [26]. That is enhancement in colour strength (K/S) by increasing the catalyst concentration could be associated with (a) reactive of the reactive dye with the methylol groups of the N-methylol finishing agent which required high catalyst concentration and (b) decomposition of N-methylol compound at high catalyst concentration which leads to amide groups with high reactivity of the reactive dye (Eq. 3,4,5,6,9).

Water/oil repellent, tactile and anticrease function as well as the dyeing properties of the simultaneous dyed/finished fabric samples are given in Table 1, the obtained data demonstrate that: (I) using of high molecular weight film-forming polymers, i.e. water and oil repellent agent in conjunction with a fibre crosslinking resin for the production of durable press fabric has been the target of enhancement crease recovery angle, water/oil repellent, handle and overall fastness properties. II) incorporation of the used remazol dyes in finishing formulation is accompanied by a significant improvement in colour yield (K/S) of the obtained dyeings along with a subsequent enhancement in their functional properties which due to the inhibition effect of pH value of trichloroacetic acid (TCAA) during heat treatment as well as inactivation effect of the used Remazol dyes.

$$\text{Cell.OH} + \text{Knittex FEL} \xrightarrow[\Delta]{H^+} \text{Cross linked cellulose} \qquad (1)$$

$$\text{HX - Ny - XH} + \text{Knittex FEL} \xrightarrow[\Delta]{H^+} \text{Crosslinked polyamide 6(Ny)} \qquad (2)$$

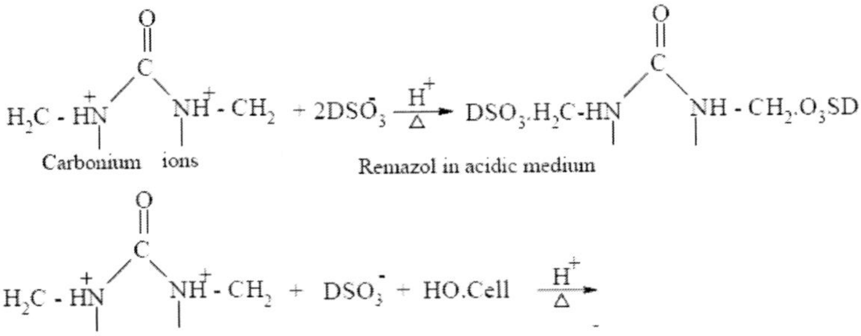

(3)

Table 1. Effect of Trichloroacetic acid (TCAA) concentration on the functional dyeings of polyamide 6/ cotton blended fabrics

Trichloro acetic acid (TCAAg/l)	K/S	Washing at 95°			Rubbing		Perspiration						Light fastness	Handle	Grease Recovery Angle				Tensile Strength		Elongation %		Repellency Rating	
							Acidic			Alkaline					Dry		Wet							
		Alt.	C	N	Dry	Alt.	Alt.	C	N	Alt.	C	N			W.D	F.D.	W.D	F.D.	W.D	F.D.	W.D	F.D.	Water	Oil
0	4.57	2	2-3	2	3	2-3	3	3	2-3	3	2-3	3	5	S	60	78	100	118	40	30	25	35	0	0
1	10.20	2-3	3-4	3	3-4	3	3-4	3-4	3	3-4	3	3-4	6	S	120	141	125	146	39	29	24	34	50	3
2	14.40	4	4	4	4	3-4	4	4-5	3-4	4	3-4	4	6-7	S	130	151	135	152	37	29	23	33	70	5
3	14.00	3-4	4	4	4	3-4	4	4	3-4	3-4	3-4	3-4	6-7	S	130	148	133	149	36	28	23	30	70	5
4	9.22	3-4	3	3-4	3-4	3	3-4	3	3-4	3-4	3	3-4	6	H	131	148	131	148	35	27	21	29	70	5

K/S: Colour strength; Alt.: Alteration; C: Staining on cotton; N: Staining on polyamide-6
W.D.: Warp direction; F.D: Weft direction; S: Soft; H: Harsh
Padding Bath Contents: Knittex® FEL (50g/l), Remazol RedX3BS (20g/l), Water and Oil Repellent (20g/l), TCAA (0-4 g/l); Wet Pick up (70%), Drying at 100°C/ 3min, Curing at 170°C/ 2 min.

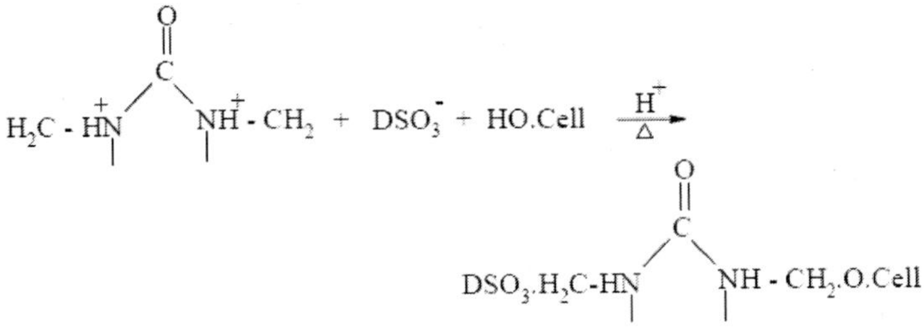

(4)

$$D.SO_2.CH_2.CH_2.OSO_3Na \xrightarrow[\Delta]{OH^-} D.SO_2.CH = CH_2 + NaHSO_4 \quad (5)$$

Reactive dye — Vinyl form

$$D.SO_2.CH_2.CH_2.OSO_3Na \xrightarrow[\Delta]{OH^+} D.SO_2.CH_2.CH_2.OS\bar{O}_3\ Na^+ \quad (6)$$

Reactive dye — Ionized form

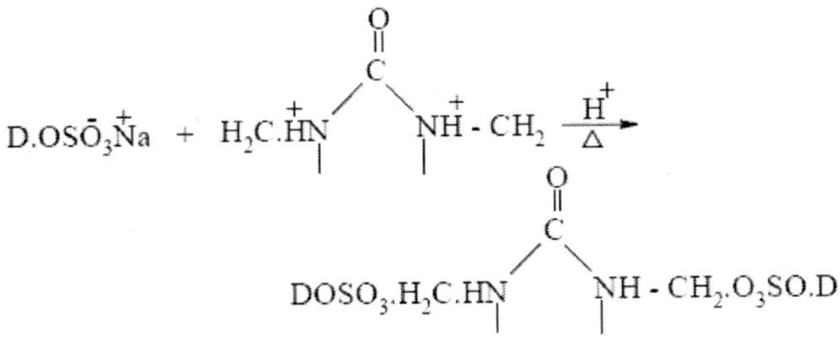

(7)

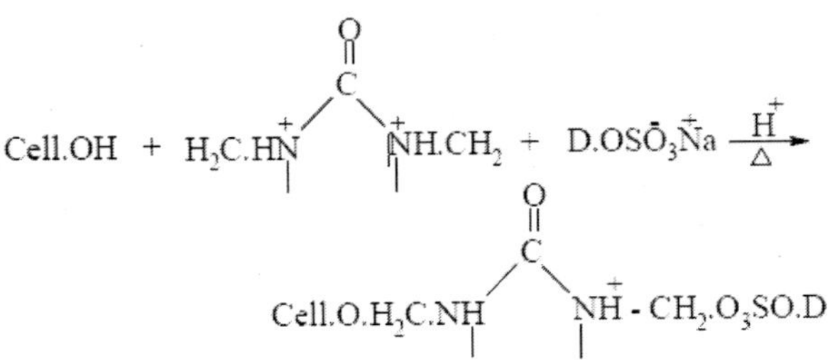

(8)

Where XH = -NH2, = NH-

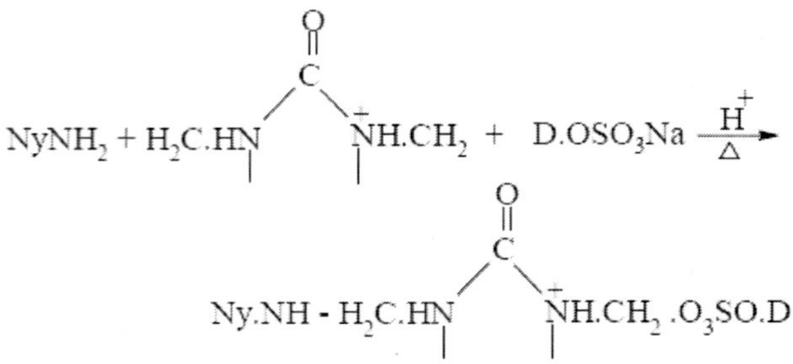

(9)

$$NyNH_2 + D.SO_2.CH_2.CH_2.OSO_3Na \xrightarrow[\Delta]{\bar{OH}} D.SO_2.CH_2.CH_2.NH.Ny \qquad (10)$$

polyamide 6 (Ny) — Vinyl sulphone

$$Cell.OH + D.SO_2.CH_2.CH_2.OSO_3Na \xrightarrow[\Delta]{\bar{OH}} D.SO_2.CH_2.CH_2.O.Cell \qquad (11)$$

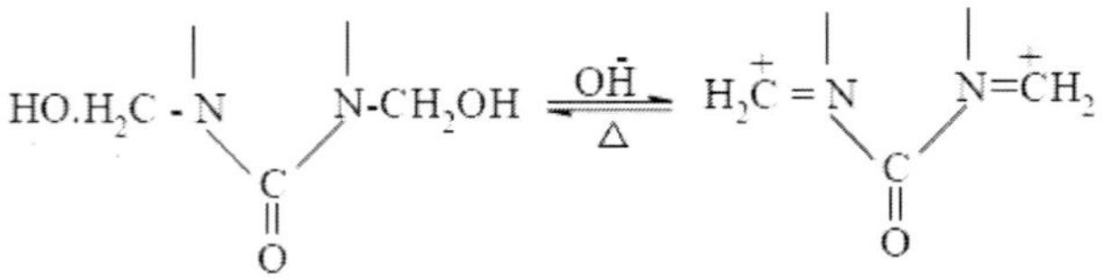

(12)

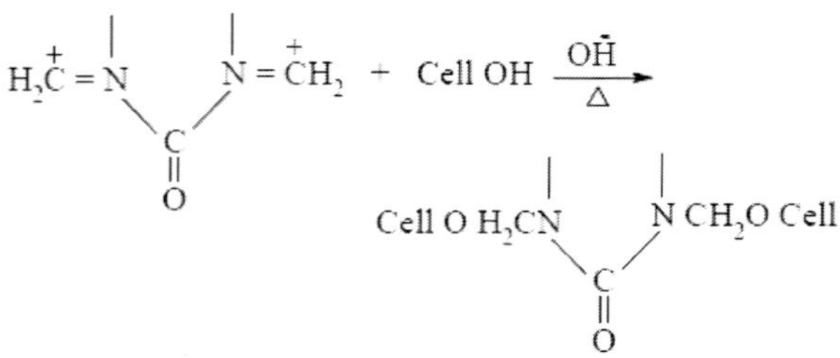

(13)

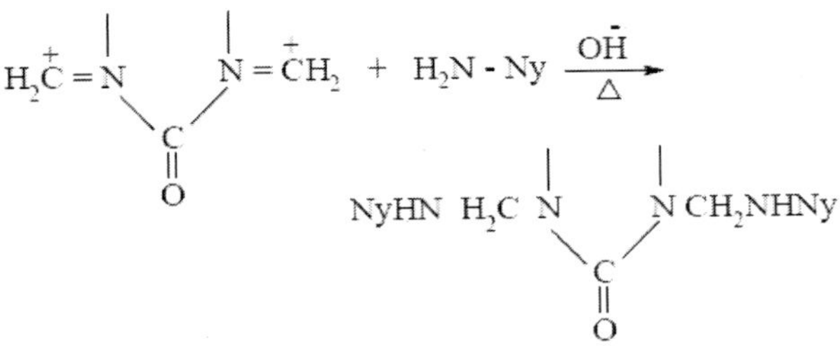

(14)

3.2. Finishing Agent Concentration

Table 2 shows the effect of increasing Knittex® FEL (0-80 g/l) as a finishing agent on the extent of crosslinking. It is clear that: (I) increasing Knittex® FEL concentration up to (60g/l) in the presence of water and oil repellent agent (20g/l), Remazol Red X3BS (20g/l), and trichloroacetic acid (2g/l) as a catalyst is accompanied by an improvement in both the crease recovery angle (CRA) and repellency rating as well as the extent of the multifunctional finishing and dyeing regardless of the used blend. (II) the extent of crosslinking as well as bifunctional properties are determined by the nature and reactivity of the substrate, it is clear from table 2 that, the Knittex ® FEL concentration increases up to 60g/l K/S values of the obtained blended fabric samples tend to increase gradually, which can be attributed to the introduction of new reactions, i.e. reaction of N-methylol finishing agent and reactive dye with polyamide 6/cotton components either in acidic or in alkaline medium due to decomposition of trichloro acetic acid (TCAA) during heat treatment, thereby improving and enhancing the extent of union reactive dye of both the polyamide 6 and Cellulosic components. A further increase in finishing agent concentration i.e., beyond 60g/l has a negative effect that can be discussed in terms of higher extent of side interaction and hydrolysis of reactive dye, thereby minimizing the availability of reactions onto finishing and dyeing/fabric matrix, irrespective of substrate used. It is also clear from (Table 2) that increasing the finishing agent concentration up to 60g/l results an increase in the extent of crosslinking expressed the crease recovery angle values, followed by a slight decrease in the aforementioned properties at higher concentration, i.e., beyond 60g/l, which could be attributed to (1) insufficient diffusion and penetration of the reactant resin into the fabric structure, (2) formation of single ended reactions rather than inter crosslinking due to variation of pH values during heat treatment.The obtained data also demonstrate that: (1) incorporation of the used reactive dyes in the finishing formulation is accompanied by a significant improvement in the colour yield K/S, overall fastness of the obtained dyeings along with a subsequent enhancement in their functional properties namely water and oil repellenty as well as easy care properties, regardless of the used dye (II) overall fastness properties are at the range of 3-4 to 4 using Knittex®FEL concentration up to 60g/l, (III) the enhancement in the functional finishing properties and overall fastness properties of the obtained dyeings most probably due to the increase in the carbonium immonium ion concentration in the proximity of the immobile active sides e.g. OH, NH_2, NH etc., in the polyamide 6/cotton structure thereby giving rise to higher extent of crosslinking [27, 28].

Thus it can be seen that using 60g/l Knittex® FEL was sufficient to obtain a Ny/Co fabric with favorable functional and performance properties.

3.3. Water and Oil Repellent Concentration

From table 3 it is clear that: within the range examined, increasing the concentration of repellent agent from zero up 30g/l is accompanied by a significant enhancement in colour strength, overall fastness properties handle, wrinkle recovery and repellent rating, beyond 30g/l concentration no improvement in the aforementioned properties, this is due to I) a direct consequence of the increased the extent of the formation of a hydrophobic/olephobic polymer film on the dyed fabric surface thereby lowering the fabric surface energy and hence repellent properties [29, 30], II) incorporation of the nominated W/O repellent in the padding bath has a positive impact on the depth of obtained dyeings along with a significant enhancement in the fastness properties ratings as a direct consequence of increasing the extent of entrapment and entanglement of the dye molecules within the crosslinked polymer film onto the fabric surface, III) W/O repellency ratings remain unaltered upon increasing the concentration from 30 to 40 g/l, V) TS and elongation decrease upon increasing the concentration of repellent agent from zero up 40g/l, due to increasing the extent of fabric cross-linking and embitterment. Further increase in the W/O repellent up to 30g/l gave soft handle, and harsh handle obtained beyond 30g/l as well as decreasing in T.S. and elongation. This is certainly attributed to the deposition and formation of a repellent film on the fibre surface thereby reducing the energy at fiber surfaces, and reducing surface inter-fibers and inter-yarn frictions.

Table 2. Effect of Knittex® FEL on bifunctional finishing and dyeing of polyamide 6/cotton (Ny/Co) blended fabrics using remazol red X3BS as a single dye

Knittex® FEL concentration g/l	K/S	Washing at 95°			Rubbing		Perspiration						Light fastness	Handle	Crease Recovery Angle				Tensile Strength		Elongation %		Repellency Rating	
							Acidic			Alkaline					Dry		Wet							
		Alt.	C	N	Dry	Alt.	Alt.	C	N	Alt.	C	N			W.D	F.D.	W.D	F.D.	W.D	F.D.	W.D	F.D.	Water	Oil
0	5.80	2-3	2	2-3	2-3	2	2-3	3	2-3	3	2-3	2-3	5	S	63	71	95	115	41	32	27	37	50	3
20	8.90	2-3	2-3	3	3	2-3	3-4	3-4	3-4	3-4	3	3	6	S	122	130	128	150	38	31	26	36	70	5
40	14.30	3	3	3-4	3-4	3	3-4	4	4	3-4	3-4	3-4	6-7	S	129	135	135	155	37	30	25	35	70	5
60	16.80	4	4	4	4	3-4	4	4	3-4	4	3-4	4	6-7	S	133	144	140	162	36	30	24	33	80	7
80	16.00	3-4	3-4	3-4	3-4	3	4	3-4	3-4	4	3	3-4	6-7	H	132	143	138	160	36	28	22	32	80	7

K/S: Colour strength Alt.: Alteration C: Staining on cotton N: Staining on polyamide-6
W.D.: Warp direction F.D: Weft direction S: Soft H: Harsh
Padding Bath Contents: Knittex® FEL (10-80g/l), Remazol RedX3BS (20g/l), W/O Repellent (20g/l), TCAA (2g/l); Wet Pick up (70%), Drying at 100°C/ 3min, Curing at 170°C min.

Table 3. Effect of water/oil repellent concentration on bath dyeing and functional properties of polyamide 6/cotton blends

W/O repellent agent (g/l)	K/S	Washing at 95°			Rubbing		Perspiration						Light fastness	Handle	Crease Recovery Angle				Tensile Strength		Elongation %		Repellency Rating	
							Acidic			Alkaline					Dry		Wet							
		Alt.	C	N	Dry	Alt.	Alt.	C	N	Alt.	C	N			W.D	F.D.	W.D	F.D.	W.D	F.D.	W.D	F.D.	Water	Oil
0	12.90	3	3	2-3	3	2-3	3	2-3	2-3	3	2-3	2-3	5	S	64	73	105	131	40	32	28	38	0	0
10	15.80	3-4	3-4	3	3-4	3	3-4	3	3	3-4	3	3	6	S	124	134	130	155	39	31	27	37	50	3
20	16.20	4	4	3-4	3-4	3-4	4	3-4	3-4	4	3-4	3-4	6-7	S	130	147	138	160	38	31	26	36	70	5
30	17.70	4-5	4-5	4	4	4	4	4	3-4	4	3-4	4	6-7	S	139	158	143	167	38	30	25	35	80	7
40	17.60	4-5	4-5	4-5	4-5	4	4-5	4	4	4-5	4	4	6-7	H	138	157	140	165	36	29	25	34	80	7

K/S: Colour strength   Alt.: Alteration   C: Staining on cotton   N: Staining on polyamide-6
W.D.: Warp direction   F.D: Weft direction   S: Soft   H: Harsh
Padding Bath Contents: Knitex® FEL (60g/l), Remazol RedX3BS (20g/l), Water/Oil Repellent (0-40 g/l), TCAA (2g/l); Wet Pick up (70%), Drying at 100°C/ 3min, Curing at 170°C/2 min.

Table 4. Effect of curing temperature on dyeing and functional properties of polyamide 6/cotton blended fabrics

Curing Temp. °C	K/S	Washing at 95°			Rubbing		Perspiration						Light fastness	Handle	Crease Recovery Angle				Tensile Strength		Elongation %		Repellency Rating	
							Acidic			Alkaline					Dry		Wet							
		Alt.	C	N	Dry	Alt.	Alt.	C	N	Alt.	C	N			W.D	F.D.	W.D	F.D.	W.D	F.D.	W.D	F.D.	Water	Oil
100°C	14.50	3-4	3	3	3-4	3	3-4	3	3	3-4	3	3-4	5	S	90	98	114	137	39	34	27	37	50	3
120°C	14.20	4	3-4	3-4	4	3-4	4	3-4	3-4	4	3-4	4	6	S	130	149	136	165	38	33	25	37	70	5
140°C	16.30	4-5	4	4	4-5	4	4-5	4	4	4	4	4-5	6-7	S	139	160	140	170	37	32	24	36	80	7
160°C	19.60	4-5	4-5	4	4-5	4	4-5	4-5	4-5	4-5	4	4-5	7	S	143	167	148	175	36	32	24	35	80	7
180°C	19.80	4-5	4-5	4-5	4-5	4	4-5	4-5	4-5	4-5	4	4-5	7	H	142	165	146	173	35	30	23	34	80	7

K/S: Colour strength Alt.: Alteration C: Staining on cotton N: Staining on polyamide-6
W.D.: Warp direction F.D: Weft direction S: Soft H: Harsh
Padding Bath Contents: Knittex® FEL (60g/l), Remazol RedX3BS (20g/l), Water/Oil Repellent (30 g/l). TCAA (2g/l); Wet Pick up (70%), Drying at 100°C/3min, Curing at 180°C/2 min.

3.4. Curing Temperature

The curing temperature is regarded as one of the most important factors in simultaneous multi-functional finishing and dyeing. Bearing this in mind, it was felt that the curing of the multi-finishing and dyeing samples must becarried out over a range of temperature to discover the optimal temperature, viz, 100°, 120°, 140°, 160° and 180°C for 2 minutes.

Crease recovery angle (CRA) and W/O repellent function as well as the dyeing properties of the simultaneously dyed/finished fabric samples are given in Table 4, the obtained data demonstrate that: I) Incorporation of the used Remazol dyes in finishing formulation is accompanied by a significant improvement in the colour yield, K/S, performance properties of the obtained dyeings along with a subsequent enhancement in their functional properties namely W/O repellent expressed by (50-80), (3-7) as well as CRA expressed by dry warp direction (90-142), dry weft direction (98-165) and wet warp direction (114-146), wet weft direction (137-173), regardless of the used dye II) fastness to washing, rubbing and perspiration of the dyeings are in the range 3 to 4-5, III) handling of the obtained dyeings through increasing the curing temperature from 100 to 160°C gave soft handle, beyond 160°C gave harsh handle.(V) Increasing the temperature from 100° to 160°C, that is the favorable effect of temperature on cross-linking cellulose chains with DMDHEU (Knittex FEL) as well as the progressive consumption of some DMDHEU in binding OLEPHOBOL®CP-U onto cotton structure through its labile hydrogen of urethane groups [31], for example, as illustrated in the following equation:

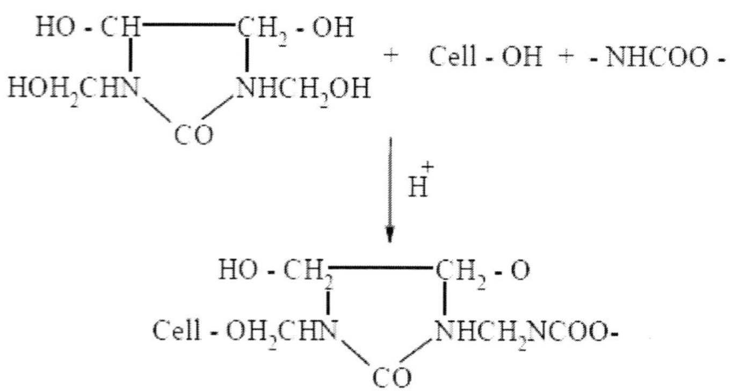

Further raising in curing temperature, i.e. beyond 160°C, has practically no or marginal effect on the aforementioned properties, which may attributed to the enhancement of side interaction and/or partial hydrolysis of substrate/ finishing agent bonds [32].

3.5. Curing Time

To avoid under-cure and over-cure, it becomes inevitable to investigate the effect of curing time on multifunctional finishing and performance properties. Curing the treated specimens were conducted at 160°C for different periods of time ranging from 1 to 5min for achieving a solid shade on polyamide-6/cotton blended samples (Table 5). An enhancement in functional finishing properties, colour strength (K/S) as well as performance properties by increasing duration of curing is somewhat expected. Reasons for this are the following a) cellulose hydroxyls of the cotton component in the blend is immobile. Their reaction with N-methylol finishing agent/ dye system would depend upon the time allowed for the molecules of this system to diffuse inside the fabrics, b) W/O repellent ratings, CRA, K/S, and performance properties attain maximum values within the time used i.e., (3 min), and prolonging it up to 5 min has no effect on these values, which suggests that most of the W/O repellent agent is located at the fabric surface with little penetration inside the fabric structure [33], c) CRA increases upon increasing time to 3 min then remain nearly constant or slightly decreases by raising the time up to 5 min., d) TS and elongation decrease by raising the time from 1 to 5 minutes, due to increasing the extent of fabric cross-linking and embitterment, e) curing at 160°C for 4 and 5 min have practically slight or no effect on the aforementioned parameteries. Obviously curing for 3 min at 160°C appears appropriate for attaining better performance properties.

3.6. Bifunctional Finishing and Dyeing Properties

W/O repellent and easy-care function as well as the dyeing properties of the simultaneously dyed / finished fabric samples and k/s as well as overall fastness properties are given in Table 6. On the other hand, the impact of the type of reactive dye on the extent of functional finishing and dyeing of fabric samples is also shown in this Table. For a given finishing and dyeing conditions, it is seen that; I) type of reactive dye has practically effect on the extent of inter-crosslinking of polyamide-6/cotton chains, expressed as CRA as well as the W/O repellent rating, II) the enhancement in the CRA, W/O repellent as well as overall fastness properties of the used substrate reflects the positive impact of the decomposition of trichloroacetic acid during curing via acidic, neutral and alkaline medium, III) fastness to washing, rubbing and perspiration of the obtained reactive dyeings are in the range of 4 to 4-5.

Table 5. Effect of curing time on dyeing and finishing properties of polyamide 6/cotton blended fabrics

Curing Time (min)	K/S	Washing at 95°			Rubbing		Perspiration						Light fastness	Handle	Crease Recovery Angle				Tensile Strength		Elongation %		Repellency Rating	
							Acidic			Alkaline					Dry		Wet							
		Alt.	C	N	Dry	Alt.	Alt.	C	N	Alt.	C	N			W.D	F.D.	W.D	F.D.	W.D	F.D.	W.D	F.D.	Water	Oil
1	14.60	4	3-4	4	3-4	3-4	3-4	3-4	4	3-4	3-4	3-4	6	S	120	141	123	143	39	33	27	37	50	3
2	18.95	4.5	4-5	4	4-5	4	4-5	4-5	4-5	4-5	4	4	4-7	S	140	168	142	172	37	32	26	36	70	5
3	20.60	4.5	4-5	4-5	4-5	4-5	4-5	4-5	4-5	4-5	4-5	4-5	7	S	149	173	152	178	36	31	24	34	80	7
4	20.70	4.5	4	4-5	4-5	4-5	4-5	4-5	4-5	4-5	4-5	4-5	7	S	147	172	151	176	36	31	24	34	80	7
5	20.75	4.5	4	4-5	4-5	4-5	4-5	4-5	4-5	4-5	4-5	4-5	7	H	145	169	150	172	35	30	23	33	80	7

K/S: Colour strength; Alt.: Alteration; C: Staining on cotton; N: Staining on polyamide-6
W.D.: Warp direction; F.D: Weft direction; S: Soft; H: Harsh

Padding Bath Contents: Knittex® FEL (60g/l), Remazol RedX3BS (20g/l), Water/Oil Repellent (30 g/l), TCAA (2g/l); Wet Pick up (70%), Drying at 100°C/ 3min, Curing at 160°C/ at different time (1-5 min).

Table 6. Technological evaluation of bifunctional finishing and dyeing properties of polyamide 6/cotton blends using different remazol dyestuffs

Dyestuff	K/S	Washing at 95°			Rubbing		Perspiration						Light fastness	Handle	Crease Recovery Angle				Tencile Strength		Elongation %		Repellency Rating	
							Acidic			Alkaline					Dry		Wet							
		Alt.	C	N	Dry	Alt.	Alt.	C	N	Alt.	C	N			W.D	F.D.	W.D	F.D.	W.D	F.D.	W.D	F.D.	Water	Oil
Remazol Red RC	19.75	4-5	4	4	4	3-4	4-5	4-5	4	4-5	4	4-5	6-7	S	147	175	150	179	39	31	30	39	70	5
Remazol Blue RC	20.55	4-5	4	4	4-5	4	4-5	4-5	4	4-5	4	4-5	6-7	S	145	173	149	178	41	30	31	43	70	5
Remazol Golden Yellow RC	13.50	4-5	4-5	4	4-5	4	4-5	4-5	4	4-5	4	4-5	7	S	148	177	149	179	42	31	31	44	80	7

K/S: Colour strength Alt.: Alteration C: Staining on cotton N: Staining on polyamide-6
W.D.: Warp direction F.D: Weft direction S: Soft H: Harsh
Padding Bath Contents: Knittex® FEL (60g/l), Remazol RedX3BS (20g/l), Water/Oil Repellent (30 g/l), TCAA (2g/l); Wet Pick up (70%), Drying at 100°C/ 3min, Curing at 160°C/ 5min.

4. CONCLUSIONS

Multi-functional finishing/reactive dyeing of polyamide-6 /cotton blends (50/50) in one stage via trichloroacetic acid (TCAA) as a catalyst has been established for imparting a specific functional properties to the above substrate.

Best results of k/s values, crease recovery angle (CRA), handle, water-oil repellency and overall fastness properties were obtained with 2 g/l of trichlroacetic acid.

The extent of bifunctional finishing/dyeing and performance properties of polyamide-6 / cotton were achieved via 60 g/l of knittex® FEL as a finishing agent.

The best multifunctional finishing/dyeing properties i.e., handle and k/s values as well as over-all fastness properties were achieved via W/O repellent agent at the concentration of 30 g/l.

The optimal curing temperature that brings about a significant improvement in the present maltifunctional finishing/dyeing properties was 160°C.

W/O repellent ratings, CRA, K/S and other functional properties attained maximum values at the optimum curing time used i.e., 3 minutes, and prolonging it up to 5 minutes has no effect.

The obtained data revealed that very good results for the over-all fastness properties of the bifunctional/dyeing polyamide-6 /cotton blends with different remazol dyes were determined at the previous optimum conditions, and they are in the range of 4 to 4-5.

REFERENCES

1. D. Hegemann, M. M. Hassain D. Balazs, Nanostructured plasma coatings to obtain multifunctional surfaces, Prog. Org. Coat. 58 (2007) 237-240.

2. L. H. Lin. C. C. Wang. C. W. Chen. K. M. Chen, Water-repellent and antibacterial activities of plasma-treated cleavable silicone surfactants on nylon fabrics surf. Coatings technical 201 (2002) 674-678.

3. S. Lee. J. S. G. Cho, Antimicrobial and blood repellent finished for cotton and nonwoven fabrics based on chitosan and fluoropolymers, Textile Res. J. 69 (1999) 104-112.

4. P. De. M. D. Sankhe S. S. Chawdhari, M. R. Mathur, UV-resist water-repellent breathable fabric as protective textiles, J. Ind. Textile 34 (2005) 209-222.

5. L. Benisek P. C. Craven Machine-washable, water and oil repellent, flame retardant wool, textile. Res. J. 12 (1980) 705-710.

6. M. M. Millard K.S. Lee. AE. Pavlath, Continuous low-temperature discharge treatment of wool yarn: a process for making wool yarn oil repellent and shrink resistant, Textile Res. J. 42 (1972) 307-313.

7. K. D. Veatch and B. M. Gatewood, AATCC Review, 2, 47-51 (2002).

8. N. A. Ibrahim, R. Refaie and A. F. Ahmed, J. Industrial Textiles, 40, 65-83 (2010).

9. P. Bajaj, "Finishing of Textile Materials", J. Appl. Polymer Sci., 83, 631-659 (2002).

10. Holme, Coloration Technologies, 123, 59-73 (2007).

11. A.Hebeish and N. A. Ibrahim, Colourage Annual, 54, (4), 41-55 (2007).

12. P. Martel, M. Weltrowski, D. Ruffin and M. Morecellet, J. Appl. Polym. Sci., 83, 1449-1456 (2002).

13. P. Horridge and S. Shan, American Dyestuff Reporter, (7), 38 (1994).

14. Intersectional Technical Paper Competition, Textile Chemist and Colorist 26, (11), 29 (1994).

15. Imada, K., Yoshida, M. Text. Chem. Color 22 (11), 18 (1990).

16. Cardamone, J. M., Bao, G Marmer, W. N. Text. Chem. Color 29 (9), 30 (1997).

17. Cardamone, J. M., Marmer, W. N., Lambert, A. H., Bardy. J. B. Text. Chem. Color 28 (11), 19 (1996).

18. Cardamone, J. M., Bao A., Frances, A; Marmer, W. N. Am. Dyest Rept. 86 (1), 41-47 (1997).

19. Park. J., Shore, J. Text. Chem. Color., Amer. Dyest Rep., 32 (1), 46 (2000).

20. Osterloh, Textile Praxis, 26 (1971) 164.

21. Givse and Stapleton, J. S. D. 91 (1975) 223.

22. Feess, Melhiand Textilber, 51 (1970) 197.

23. Kaiser, Chemiefasern/ Textilindustrie, 26, 128 (1976) E63.

24. Kubelka, P. Munk, F. Texhn. Z. Physik 1931, 12, 593.

25. H. Peterson "Chemical After-treatment of Textiles" Edited by H. Mark, H. S. Wooding and S. M. Atlas, John Wiley & Sons, Inc., (1971).

26. Hebeish, E. A. Allam, A. A. Hamsa and M. Kamel., Kolorisztikat Ertesito, 11.12, 308 (1976).

27. W. K. Lee, and K. M. Sin, Textile Asia, 22 (12), 86 (1991).

28. P. Xiaojum, W. Jitao, and S. Jie, J. S. D. C., 109, 159 (1993).

29. N. A. Ibrahim, A. Amr, B. M. Eid and Z. M. El Sayed, "Innovative Multifunctional Treatments of Ligno. Cellulosic Jute Fabric" Carbohydrate Polymers, 82, 1198-1204 (2010).

30. M. H. Abo Shosha. Z. H. El-Hilw, A. A. Aly and A. Amer, "New Textile Water Repellent Based on Reaction of Toluene 2,4 Düsocyanate with Stearyl Alcohol" AATCC Review, 9, 38-42 (2009).

31. Abo. Shosha, M. H., Ibrahim, N.A. "Water soluble polyurethane as finishing additives is easy care finishing of cotton fabrics, Sci. International (Lahore), 1992, 4 (2), 163.

32. Haggag, K, Ebrahim, N. A. J. Text. 1991, 32. 9.

33. Adams, N. E. Detergent action and its relation to wetting and emulsification. J. Soc. Dyers and Color. 1937, 53, 121.

Chapter 2

Effect of Different Dyeing Parameters on Color Strength & Fastness Properties of Cotton-Elastane (CE) and Lyocell-Elastane (LE) Knit Fabric

Md. Abdus Shahid[1], Md. Ismail Hossain[2], Delwar Hossain[1], Ayub Ali[3]

[1]Dhaka University of Engineering & Technology (DUET), Department of Textile Engineering, Gazipur, Bangladesh

[2]University of Malaya (UM), Department of Mechanical Engineering, Kuala Lumpur, Malaysia

[3]BGMEA University of Fashion and Technology (BUFT), Department of Apparel Manufacture & Technology, Dhaka, Bangladesh

ABSTRACT

Color fastness and color strength are very crucial facts in case of colored textile material which are intimately related to various dyeing parameters viz. concentration of dye, salt, alkali and dyeing temperature. In the present work, the effect of concentration of reactive dye (1%, 3%, 5%, 7% & 9%), salt (20g/l, 30g/l & 40g/l) and alkali (5g/l, 8g/l & 10g/l), and the effect of dyeing temperature (40°C, 60°C & 80°C) on color fastness and color strength properties of cotton-elastane (CE) and lyocell-elastane (LE) fabrics were studied remaining the constant of dyeing duration time at 45 minutes. The results were compared for variation in fastness values and color strength under the same conditions as being employed in textile industry. Some of the factors behave linearly and some are not. Result reflects that the color fastness values to dry and wet rubbing are more at low dye concentration, low salt and alkali concentration and at low temperatures for both cotton-elastane and lyocell-elastane fabrics which reaches up to 4-5 to 5. But in case of washing fastness property, both fabrics reflect a fixed value i.e.4-5 irrespective of dye concentration, salt and alkali concentration, dyeing time and temperature. On the other hand, the color strength results evinced that temperature has more influence on LE than CE for acquiring higher color strength. LE fabric exhibited better dye ability with higher color uptake compared to CE fabric. In addition the concentrations of salt and alkali have almost linear proportional effect on both CE and LE within a range.

KEYWORDS

Cotton-Elastane, Lyocell-Elastene, Color Fastness, Color Strength, Reactive Dye, Salt, Alkali

1. INTRODUCTION

The generations of regenerated cellulose fibers, such as viscose, modal and lyocell fibers are among the most important fibers from the point of textile and environmental aspects due to the natural structures and properties [1]. Lyocell is the generic name for a regenerated cellulosic fiber obtained by spinning of dissolved wood pulp in an organic solvent. These regenerated cellulosic fibers are produced by a more environmentally friendly procedure where cellulose is dissolved in the organic solvent N-methyl morpholine-N- oxide (NMMO). Regeneratedcellulosic yarns with various chemical reagents had been the subject of research for several years for NMMO or lyocell fibers, produced by the amine oxide process [2, 3] are solvent spun. More than 99% of the solvent is recycled within the process, making the fiber production ecofriendly. Very small quantities of NMMO emitted via waste water are readily degraded in the biological waste water treatment plants. NMMO with its favorable toxicological properties has been found to be less toxic than ethanol. The solvent spinning technique so adopted is an environmentally responsible process [4]. This manufactured fiber shows round cross-sections and a smooth surface. Their high degree of crystallinity, molecular orientation and higher molecular weight causes special properties such as high wet fiber strength and comparatively low elongation, leading to the remarkable dimensional stability of fabrics produced from them. Lyocell fiber's degree of crystallinity is 16 % higher than that of modal fibers and much higher (43%) compared to viscose fibers. The fibrils in lyocell fibers are thinner than in modaland viscose fibers and molecular orientation is nearly the same (orientation function of lyocell fibers is 3% higher) as in modal fibers but much higher compared to viscose standard fibers (18%). Lenz and Schurz [5] determined the size of crystalline regions of regenerated cellulose fibers: length 12 - 14 nm, width 8 - 10 nm, and thickness 3 - 4 nm. The broad plane of the platelet-shaped crystallites is situated parallel to the 101 - lattice plane. The crystallites form strands with a length of 150 nm - 550 nm. They are partly bundled up in clusters with diameters of 30 - 60 nm, partly separated by less dense spacing regions. Since lyocell is cellulose based fiber and therefore can be dyed with any class of dyestuff suitable for other cellulosic fiber. On the other hand cotton fibers are widely applied in textile industry due to its excellent properties of hygroscopicity, air permeability, biodegradability, no static electricity, etc. The dyeing of these fibers are generally done with reactive dyes due to its brilliancy, variety of hue, high wet fastness, convenient usage and high applicability. These reactive dyes contain a reactive group, either a halo-heterocyclic or an activated double bond, that, when applied to a fiber in an alkaline dye bath, forms a chemical bond with hydroxyl group on the cellulosic fiber [6]. In recent years, reactive dyes maintain the largest annual consumption in the world among the dyes used for which establishes its important status in the dye manufacture industry. But some problems, such as low dye utilization, large amount of electrolyte used and high volume of wastewater discharged, always exist in the application of reactive dyes. The dyeing of one kilogram of cotton with reactive dyes demands from 70 to 150 litre water, 0.6-0.8 Kg NaC1 and from 30 to 60 g dyestuffs [7]. Spandex/elastane is an elastomeric which is always blended with other natural and synthetic fibers such as cotton, wool, silk, and linen. Different types of spandex material exist on the markets with different counts, this impose engineering the use of every type at different spun materials and fabric construction. The percentage of spandex used in the knitting industry depends on the fabric stretch required and fabric properties [8]. The reactive dyes have higher natural affinity for LE fabric than for CE fabric because of its structure as a result it shows higher K/S value in comparison with CE fabric. Hence, reactive dyes with good migration properties will be best suited for LE fabric providing level dyeing effect. Thus, the electrolyte recommendations can also be significantly reduced, as compared to the dyeing to the same depth of shade on CE fabric. Consequently, there is reduction in chemical cost, handling and total dissolved solids. Lower electrolyte also facilitates quicker removal of unfixed hydrolysed dye from the fiber in washing stage after dyeing, thus reducing the process cycle time and hence the cost [9]. Traditionally disperse dye was used to dye spandex/ superfine polyester fabric and some auxiliaries were also approached for dyeing of elastic fiber with disperse dye [10, 11]. Later on reactive dyes are in fact seen as the major dye class used on LE fabric and CE fabric. However, it is revealed that cotton with less elastane percentage shows better fastness property in case of reactive dyeing [12]. There are several studies done on the optimization of process parameter for obtaining improved dyeing behavior. The process parameters have a great influence on the dyeing performance [13, 14]. Some pretreatment process and treatments give better result in colorfastness property of polyester-spandex fabric [15]. In case of dyeing the CE fabric, ways of getting better color fastness property has been explained by using auxiliaries (HYS-422 auxiliaries) [16]. However, the effect of variation of different conditions (temperature and dye concentration) and auxiliaries (salt and alkali) have not been found during researching by the authors. Furthermore another most eco-friendly fiber LE fabric dyed with reactive dye and the influence the process parameters along with their comparison with CE in respect to color fastness and color strength requires clarification. The aim of this study was to investigate and compare the effect of temperature, amount of alkali, amount of salt and dye concentration ondyability characteristics especially on color fastness (rubbing and wash) and color strength of CE and LEknit fabric.

2. MATERIALS AND METHODS

2.1. Procedure of CE and LE Knit Fabric Manufacturing

In this investigation, single jersey cotton-elastane (CE) and lyocell-elastane (LE) knit fabrics were manufactured by single jersey knitting machine having 32 inches diameter, 24 gauges (needles/inch) and 104 yarn feeders (Brand: Fukuhara, Model: MXC-S3.2RE, Origin: Japan). CE knit fabric was made from 95% cotton & 5% elastane yarn and LE grey fabric was made from 95% lyocell & 5% elastene yarn. The yarn count was 32^s/1for cotton, 18 tex for lyocell, 30 denier for lycra. 100% cotton yarn was sourced from Square Spinning mills Ltd. Bangladesh, lyocellyarn (Brand: Tencell) from Austria, and elastence yarn (Brand: Texlon) from Korea.

2.2. Pre-treatment & Dyeing Procedure

Both types of grey fabrics were scoured and bleached using single bath. Recipe for CE fabric was NaOH @1.2 g/l, H_2O_2@2.5 g/l at 98°C for 60 minutes during scouring & bleaching. On the other hand, recipe for LE fabric was Na_2CO_3@4 g/l, H_2O_2@2.5 g/l at 80°C for 60 minutes. Single bath scouring and bleaching process was done in a laboratory dyeing machine (Brand: Starlet, Model: DL-6000, Origin: Korea) maintaining the material & liquor ratio of 1:6. Average grams per square meter (GSM) of scoured & bleached fabric were 185. All pre-treated CE and LE knit fabric samples were dyed using exhaust dyeing methods with reactive dye (Remazol Blue RR) in a laboratory dyeing machine (Brand: Starlet, Model: DL-6000+, Origin: Korea) according to Table 1 at material and liquor ratio of 1:8. Forty five experiments were carried out altering the dyeing parameters viz. concentration of dye, salt & alkali and temperature remaining the constant of dyeing duration time at 45 minutes that is showed in Table 1. Table 1 shows that five different level of dye concentration namely 1%, 3%, 5%, 7% and 9% with different salt & alkali concentration and dyeing temperature were used for preparing the samples during this study. After dyeing all samples were cold rinsed and then hot washed by washing machine (Brand: LG, Model: WD-1203 CDP, Origin: Korea), at 90°C for 10 minutes. Next, the samples were dried by tumble dryer machine (Brand: SDL, Origin: UK).and conditioned for 2 hours at (65±2) % RH (relative humidity) and (20±2) °C temperature.

2.3. Measurement of Color Strength and Color Fastness Properties

Reflectance values of all dyed samples were measured using spectrophotometer (Brand: Datacolor Spectrum, Origin: USA, Model: 650™), in a visible region with wavelength of 650nm. The color strength (K/S) was calculated using the Kubelka-Munk equation:

$$\frac{K}{S} = \frac{(1-R)^2}{2R}$$

Table 1. Experimental design considering the dyeing parameters

Exp. no.	DC %	Dye temp (^{0}C)	Time (min)	SC (g/l)	AC (g/l)
1	1	40	45	20	5
2	1	60	45	20	5
3	1	80	45	20	5
4	3	40	45	20	5
5	3	60	45	20	5
6	3	80	45	20	5
7	5	40	45	20	5
8	5	60	45	20	5
9	5	80	45	20	5
10	7	40	45	20	5
11	7	60	45	20	5
12	7	80	45	20	5
13	9	40	45	20	5
14	9	60	45	20	5
15	9	80	45	20	5
16	1	40	45	30	8
17	1	60	45	30	8
18	1	80	45	30	8
19	3	40	45	30	8
20	3	60	45	30	8
21	3	80	45	30	8
22	5	40	45	30	8
23	5	60	45	30	8
24	5	80	45	30	8
25	7	40	45	30	8
26	7	60	45	30	8
27	7	80	45	30	8
28	9	40	45	30	8
29	9	60	45	30	8
30	9	80	45	30	8
31	1	40	45	40	10
32	1	60	45	40	10
33	1	80	45	40	10
34	3	40	45	40	10
35	3	60	45	40	10
36	3	80	45	40	10
37	5	40	45	40	10
38	5	60	45	40	10
39	5	80	45	40	10
40	7	40	45	40	10
41	7	60	45	40	10
42	7	80	45	40	10
43	9	40	45	40	10
44	9	60	45	40	10
45	9	80	45	40	10

*SC=Salt concentration, AC=Alkali concentration, DC%=Dye concentration %

Color fastness properties of all dyed samples were determined using the crock meter (Brand: SDL, Origin: UK) and multi-fiber. The color fastness values were assessed using grey scale both for color fastness to rubbing and color fastness to washing. Rubbing fastness properties were examined by applying BS EN ISO-105-X12:2002 method and color fastness to washing were examined applying BS EN ISO 105 CO6:2010 method.

3. RESULT AND DISCUSSION

3.1. Effect of Temperature on Color Strength and Fastness

3.1.1. Effect of Temperature on Color Strength

K/S value in Figure 1(a), Figure 2(a) and Figure 3(a) reflects that dyeing temperature has a great impact on dye uptake for both fabrics. A dramatic change occurs in fiber polymer system when temperature is applied as a medium of energy. This scenario is best viewed from the plotted curves. As the temperature increases, the molecular structure becomes open which facilitate the dye uptake and hence the higher K/S value is obtained. An interesting relation was obtained when the temperature was increased at different level of dye, salt and alkali concentration. In case of cotton-elastane, the results show that K/S value of dyed fabric increased gradually with the increase of temperature from 40°C to 60°C. In fig 3, at 60 °C for cotton-elastane shows optimum dye uptake but the K/S value in the plotted figure shows that at 80°C the dye uptake results getting down which may be due to diffusion of dye from the core of the fiber. At 1%, 3% and 5% the results shows near parallel effect. On 7% dye it shows slightly higher differences but at 9% dye concentration with 60°C and 80°C shows less difference in results whereas at 40°C it shows declined value. Again, for LE fabric samples, the plotted curves show different action. At 1%, 5% and 7% of dye concentration of 40°C and 80°C shows small value difference. But high level of dye uptake was shown at 60°C for LE. In addition, at 9% dye concentration with 80°C the LE fabrics show a highest peak but decreasing rate of dye uptake at 40°C.The reason for this is obvious and indicates of the effects of different structure for cotton and lyocell. Due to higher degree of crystalinity in the polymer structure of CE to LE causes less dye uptake of CE fabrics.

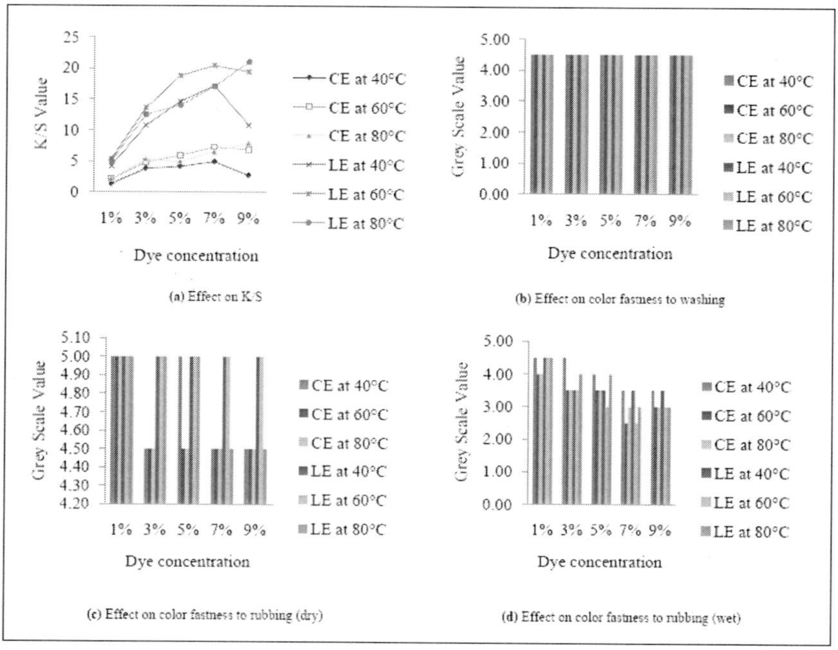

Figure 1. Effect of dye concentration and temperature at 20 g/l salt & 5g/l alkali

3.1.2. Effect of Temperature on Color Fastness

As the temperature increases, the depth of shade also increases. As a result, when the dyed samples are rubbed against an object, more dye molecule comes out from the fabric surface hence the value become less which means rubbing fastness property is not good. From the Figure 1 (c & d) and Figure 2 (c & d) for both fabrics, the rubbing fastness values are higher at a particular temperature.

In some cases, the rubbing fastness value increases with temperature which may be due to good fixation of the dye molecule at that temperature to these particular fiber. However, the effects of temperature on washing fastness properties of the dye molecule are not as visible as in case of rubbing fastness. Figure.3.b reflects that irrespective of dyeing temperature and fabric type, color staining values to multi-fiber (DW) are same at all temperature for both fabrics.

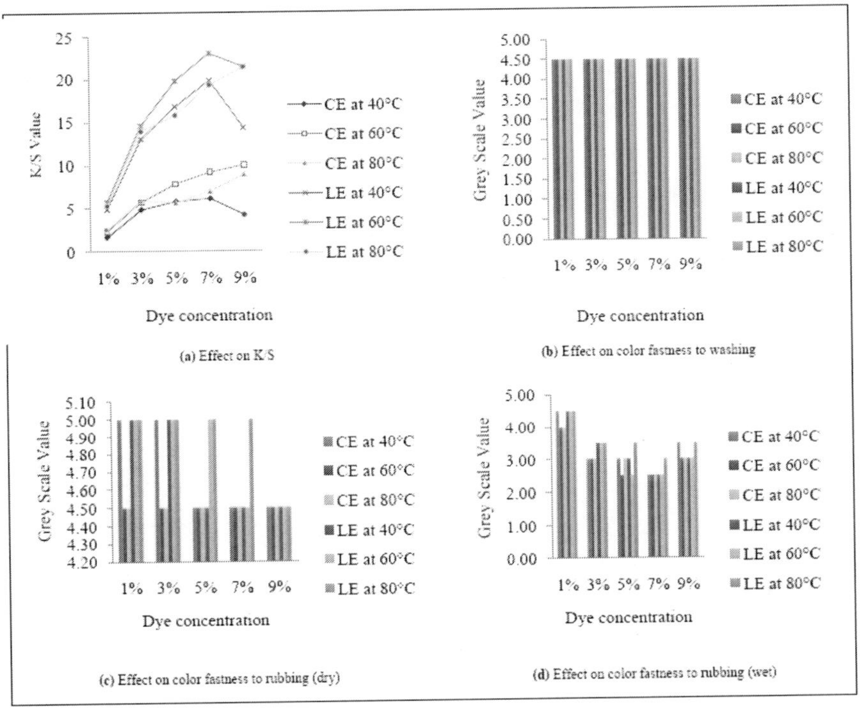

Figure 2. Effect of dye concentration and temperature at 30 g/l salt & 8g/l alkali on color strength & fastness properties

3.2. Effect of Salt and Alkali on Color Strength and Fastness

3.2.1. Effect of Salt and Alkali on Color Strength

The plotted curves of Figure 1(a), Figure 2(a) and Figure 3(a) indicate that with the increased amount of salts and alkali the dye uptake also increased. The plotted data shows a gradually increased curve with linear relation. The highest value of dye uptake was observed 11.85 for CE and 22.6 for LE with the amount of 40g/l salt and 10g/l of salt and alkali concentration respectively. The least value of salt and alkali concentration i.e. 20 g/l and 5 g/l shows the minimum K/S value for CE and LE fabrics which mean that this concentration has less impact at the values of dye uptake for fabrics. However, the effect of salt and alkali concentration on dye uptake is also affected by other factors such as-dye concentration, dyeing temperature and time etc.

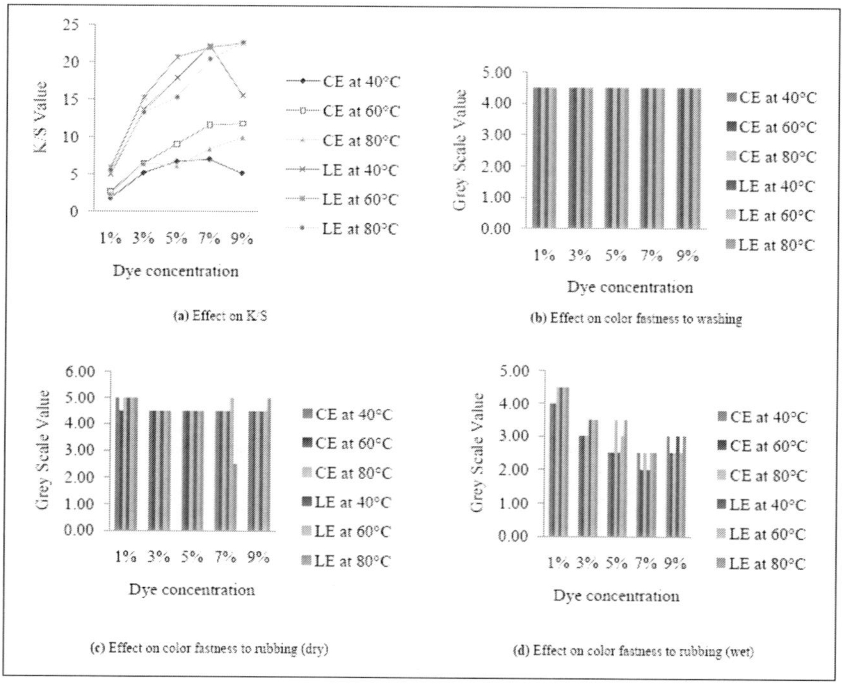

Figure 3. Effect of dye concentration and temperature at 40g/l salt & 10g/l alkalion color strength & fastness properties

3.2.2. Effect of Salt and Alkali on Color Fastness

The plotted curves in Figure 1(d) & Figure 2(d) indicate that with the increased amount of salts and alkali, the dye uptake also increased for both fabrics as a result more dye molecule comes out during wet rubbing. The higher value of grey scale in Figure 2 (c & d) indicates that dyed sample contained less dye molecule at low concentration of salt and alkali. At low concentration of alkali and salt, the grey scale value for dry and wet rubbing reached upto 4.5-5 which means good rubbing fastness. However, salt and alkali concentration has no significant effect on wash fastness properties of the both samples as shown in Figure 3(c).

3.3. Effect of Dye Concentration on Color Strength and Fastness

3.3.1. Effect of Dye Concentration on Color Strength

K/S value in Figure 1(a), Figure 2(a) and Figure 3(a) reflects that dye concentration has an unavoidable impact on dye uptake for both fabrics. Generally as the dye concentration increases, the availability of the dye molecules in the dye bath also increases as a result more dye molecule can attach to the fiber and hence the higher K/S value is obtained. In case of CE fabric, the results show that K/S value of dyed fabric increased gradually as the dye concentration shifted from 1 to 9% along with others parameters i.e. temperature, salt and alkali concentration with some exceptions. Certainly, the maximum K/S values (11.85) for CE and 22.68 for LE fabric were achieved at 9% dye concentration with some deviations in temperature. The deviation in K/S value at dye concentration 7% and 9% for both fabrics is similar than other values.

3.3.2. Effect of Dye Concentration on Color Fastness

The effect of dye concentration on rubbing fastness properties of the dyed samples for both fabrics is more significant than other parameters. Generally as the dye concentration increases, the availability of the dye molecules in the dye bath also increases as a result more dye molecule can attach to the fiber and hence depth of shade increases. From Figure 1(c & d) and Figure 2(c & d), the result shows that as the dye percentage shift from 1 to 9%, the grey scale value declines because dye molecule from dyed sample easily came out during rubbing. On the other hand, irrespective of dye concentration for both fabrics in Figure 3(b), it can say that color staining values are less. This may be happen due to two factors; for low dye concentration few dye molecules came out during washing as a result staining of multi-fiber was less and for higher dye concentration, dye molecules may be fixed strongly which increased the resistance to washing of the dye molecules.

4. CONCLUSIONS

A correlative study was made on color fastness properties &color strength of cotton-elastane (CE) and lyocell-elastane (LE) for different dyeing parameters. The color fastness values for both fabrics show maximum value i.e.4-5 to 5 at a particular temperature with some exceptions. The fastness values of lyocell-elastane fabrics is more in comparison with cotton-elastane fabrics for dry and wet rubbing which may be due to its high fixation rate. In case of washing fastness properties both fabrics show any differences in their values of 4-5. Therefore, it can be stated that rubbing fastness properties is inversely related with dye concentration excluding washing fastness with respective salt and alkali concentration, dyeing time and temperature. In case of color strength assessment, results showed that highest K/S value for CE fabric was obtained at dye concentration 9%, salt concentration 40g/l, alkali concentration 10 g/l and dyeing temperature 60°C whereas highest K/S value was recorded at dyeing temperature of 80°C with same concentration of salt & alkali in LE fabric. In case of CE fabric, the maximum K/S value has reached up to 11.85 and for LE fabric, K/S was 23.03.

ACKNOWLEDGEMENTS

The authors are thankful to the management of Alim Knit (BD) Ltd for providing the laboratory facilities for this study.

REFERENCES

1. Albercht. W at al. “Lyocellfibres”, Chem Fibers Int, 47, (1997): 289-304.

2. Krassig, H., and Structure Cellulose. "Accessibility and Reactivity Gordon and Breach Science Publishers." (1992).

3. Stana-Kleinschek, K., Ribitsch, V., Kreže, T., Sfiligoj-Smole, M., & Peršin, Z. Correlation of regenerated cellulose fibres morphology and surface free energy components. *Lenzinger Berichte*, *82*(1), (2003):83-95.

4. Chavan, R. B., and A. K. Patra. "Development and processing of lyocell." *Indian journal of fiber and textile research" 29.4 (2004): 483.*

5. Lenz, J., J. Schurz, and E. Wrentschur. "Comparative characterization of solvent spun cellulose and high wet modulus viscose fibers by their long periods." *Actapolymerica* 43.6 (1992): 307-312.

6. Rattee, I. D. "Reactive dyes for cellulose 1953–1983." *Review of Progress in Coloration and Related Topics*14.1 (1984): 50-57.

7. Allegre, C., et al. "Treatment and reuse of reactive dye effluents." *Journal of Membrane Science* 269.1 (2006): 15-34.

8. Haji, M. A.Properties of cotton/spandex single jersey fabrics. *The Indian Textile Journal* (2013).

9. Taylor, James M., Michael J. Bradbury, and Stephen Moorhouse. "Applied Technology-Dyeing Tencel and Tencel A100 with poly-functional reactive dyes." *AATCC Review-American Association of Textile Chemists and Colorists* 1.10 (2001): 21-24.

10. Jiangang, L. Dyeing of Spandex/Superfine Polyester Knitgoods. *Dyeing and Finishing, 7,* (2003):006.

11. Hong_fei, Q. I. A. N., & Xin_yuan, S. O. N. G.. *"The Effect of Auxiliaries on Dyeing Spandex With Disperse Dyes. Textile Auxiliaries", 2,* (2001): 007.

12. GAO, X. H., & SONG, X. Y. Research on the dyeing and finishing processes of the new cotton/spandex elastic fabrics. *Textile Auxiliaries, 6,* (2005): 010.

13. Kabir, S. F., Iqbal, M. I., Sikdar, P. P., Rahman, M. M., & Akhter, S. *"Optimization of Parameters of Cotton Fabric Whiteness". European Scientific Journal, 10*(36) (2014).

14. Shaohua, L. Pretreatment Parameters and Dyestuffs Optimization before Ink Jet Printing. *Dyeing and Finishing,12,* (2001):001.

15. QIAN, Q. F., ZHANG, J. F., ZHANG, J. G., & JIN, X. H. Wet processing of polyester/spandex fabric. *Dyeing & Finishing, 7,* (2013):005.

16. LI, X. J., LI, Y. Y., & MA, D. F. Technology Practice of Dyeing and Finishing Cotton/Spandex Undershirt Cloth.*Shandong Textile Science & Technology, 2,* (2004):006.

Chapter 3

Textile Dyes: Dyeing Process and Environmental Impact

Farah Maria Drumond Chequer[1], Gisele Augusto Rodrigues de Oliveira[1], Elisa Raquel Anastácio Ferraz[1], Juliano Carvalho Cardoso[2], Maria Valnice Boldrin Zanoni[2] and Danielle Palma de Oliveira[1]

[1] USP, Department of Clinical, Toxicological and Bromatological Analyses, Faculty of Pharmaceutical Sciences at Ribeirão Preto, University of São Paulo, Ribeirão Preto, SP, Brazil

[2] UNESP, Department of Analytical Chemistry, Institute of Chemistry at Araraquara, State of São Paulo University Júlio de Mesquita Filho, Araraquara, SP, Brazil

1. INTRODUCTION

Dyes may be defined as substances that, when applied to a substrate provide color by a process that alters, at least temporarily, any crystal structure of the colored substances [1,2]. Such substances with considerable coloring capacity are widely employed in the textile, pharmaceutical, food, cosmetics, plastics, photographic and paper industries [3,4]. The dyes can adhere to compatible surfaces by solution, by forming covalent bond or complexes with salts or metals, by physical adsorption or by mechanical retention [1,2]. Dyes are classified according to their application and chemical structure, and are composed of a group of atoms known as chromophores, responsible for the dye color. These chromophore-containing centers are based on diverse functional groups, such as azo, anthraquinone, methine, nitro, arilmethane, carbonyl and others. In addition, electrons withdrawing or donating substituents so as to generate or intensify the color of the chromophores are denominated as auxochromes. The most common auxochromes are amine, carboxyl, sulfonate and hydroxyl [5-7].

It is estimated that over 10,000 different dyes and pigments are used industrially and over 7×10^5 tons of synthetic dyes are annually produced worldwide [3,8,9]. Textile materials can be dyed using batch, continuous or semi-continuous processes. The kind of process used depends on many characteristics including type of material as such fiber, yarn, fabric, fabric construction and garment, as also the generic type of fiber, size of dye lots and quality requirements in the dyed fabric. Among these processes, the batch process is the most common method used to dye textile materials [10].

In the textile industry, up to 200,000 tons of these dyes are lost to effluents every year during the dyeing and finishing operations, due to the inefficiency of the dyeing process [9]. Unfortunately, most of these dyes escape conventional wastewater treatment processes and persist in the environment as a result of their high stability to light, temperature, water, detergents, chemicals, soap and other parameters such as bleach and perspiration [11]. In addition, anti-microbial agents resistant to biological degradation are frequently used in the manufacture of textiles, particularly for natural fibers such as cotton [11,12]. The synthetic origin and complex aromatic structure of these agents make them more recalcitrant to biodegradation [13,14]. However, environmental legislation obliges industries to eliminate color from their dye-containing effluents, before disposal into water bodies [9,12].

The textile industry consumes a substantial amount of water in its manufacturing processes used mainly in the dyeing and finishing operations of the plants. The wastewater from textile plants is classified as the most polluting of all the industrial sectors, considering the volume generated as well as the effluent composition [15-17]. In addition, the increased demand for textile products and the proportional increase in their production, and the use of synthetic dyes have together contributed to dye wastewater becoming one of the substantial sources of severe pollution problems in current times [6,9].

Textile wastewaters are characterized by extreme fluctuations in many parameters such as chemical oxygen demand (COD), biochemical oxygen demand (BOD), pH, color and salinity. The composition of the wastewater will depend on the different organic-based compounds, chemicals and dyes used in the dry and wet-processing steps [6,18]. Recalcitrant organic, colored, toxicant, surfactant and chlorinated compounds and salts are the main pollutants in textile effluents [17].

In addition, the effects caused by other pollutants in textile wastewater, and the presence of very small amounts of dyes (<1 mg/L for some dyes) in the water, which are nevertheless highly visible, seriously affects the aesthetic quality and transparency of water bodies such as lakes, rivers and others, leading to damage to the aquatic environment [19,20].

During the dyeing process it has been estimated that the losses of colorants to the environment can reach 10–50% [13,14,17,21,22]. It is noteworthy that some dyes are highly toxic and mutagenic, and also decrease light penetration and photosynthetic activity, causing oxygen deficiency and limiting downstream beneficial uses such as recreation, drinking water and irrigation [13,14,23]

With respect to the number and production volumes, azo dyes are the largest group of colorants, constituting 60-70% of all organic dyes produced in the world [2,24]. The success of azo dyes is due to the their ease and cost effectiveness for synthesis as compared to natural dyes, and also their great structural diversity, high molar extinction coefficient, and medium-to-high fastness properties in relation to light as well as to wetness [2,25]. They have a wide range of applications in the textile, pharmaceutical and cosmetic industries, and are also used in food, paper, leather and paints [26,27]. However, some azo dyes can show toxic effects, especially carcinogenic and mutagenic events [27,28].

The toxic effects of the azo dyes may result from the direct action of the agent itself or of the aryl amine derivatives generated during reductive biotransformation of the azo bond [22]. The azo dyes entering the body by ingestion can be metabolized to aromatic amines by the azoreductases of intestinal microorganisms. If the dyes are nitro, they can be metabolized by the nitroredutases produced by the same microorganisms [29]. Mammalian liver enzymes and other organizations may also catalyze the reductive cleavage of the azo bond and the nitroreduction of the nitro group. In both cases, if N-hydroxylamines are formed, these compounds are capable of causing DNA damage [29, 30].

One of the most difficult tasks confronted by the wastewater treatment plants of textile industries is the removal of the color of these compounds, mainly because dyes and pigments are designed to resist biodegradation, such that they remain in the environment for a long period of time. For example, the half-life of the hydrolyzed dye Reactive Blue 19 is about 46 years at pH 7 and 25°C [31,32].

Carneiro et al. (2010) designed and optimized an accurate and sensitive analytical method for monitoring the dyes C.I. Disperse Blue 373 (DB373), C.I. Disperse Orange 37 (DO37) and C.I. Disperse Violet 93 (DV93) in environmental samples. This investigation showed that DB373, DO37 and DV93 were present in both untreated river water and drinking water, indicating that the effluent treatment (pre-chlorination, flocculation, coagulation and flotation) generally used by drinking water treatment plants, was not entirely effective in removing these dyes. This study was confirmed by the mutagenic activity detected in these wastewaters [33].

In this context, and considering the importance of colored products in present day societies, it is of relevance to optimize the coloring process with the objective of reducing the environmental impact of the textile industry. For this purpose, liposomes could be used to carry several encapsulated dyes, and hence improve the mechanical properties of textile products, resulting in better wash fastness properties and reducing the process temperature, thus economizing energy [34]. Another way is to use ultrasonic energy, studied with the objectives of improving dye productivity and washing fastness, and reducing both energy costs and water consumption [35].

Considering the fact that the textile dyeing process is recognized as one of the most environmentally unfriendly industrial processes, it is of extreme importance to understand the critical points of the dyeing process so as to find alternative, eco-friendly methods.

2. DYEING PROCESS

The dyeing process is one of the key factors in the successful trading of textile products. In addition to the design and beautiful color, the consumer usually looks for some basic product characteristics, such as good fixation with respect to light, perspiration and washing, both initially and after prolonged use. To ensure these properties, the substances that give color to the fiber must show high affinity, uniform color, resistance to fading, and be economically feasible [36].

Modern dyeing technology consists of several steps selected according to the nature of the fiber and properties of the dyes and pigments for use in fabrics, such as chemical structure, classification, commercial availability, fixing properties compatible with the target material to be dyed, economic considerations and many others [36].

Dyeing methods have not changed much with time. Basically water is used to clean, dye and apply auxiliary chemicals to the fabrics, and also to rinse the treated fibers or fabrics [37]. The dyeing process involves three steps: preparation, dyeing and finishing, as follows:

Preparation is the step in which unwanted impurities are removed from the fabrics before dyeing. This can be carried out by cleaning with aqueous alkaline substances and detergents or by applying enzymes. Many fabrics are bleached with hydrogen peroxide or chlorine-containing compounds in order to remove their natural color, and if the fabric is to be sold white and not dyed, optical brightening agents are added [37].

Dyeing is the aqueous application of color to the textile substrates, mainly using synthetic organic dyes and frequently at elevated temperatures and pressures in some of the steps [37,38]. It is important to point out that there is no dye which dyes all existing fibers and no fiber which can be dyed by all known dyes [39]. During this step, the dyes and chemical aids such as surfactants, acids, alkali/bases, electrolytes, carriers, leveling agents, promoting agents, chelating agents, emulsifying oils, softening agents etc [23,38] are applied to the textile to get a uniform depth of color with the color fastness properties suitable for the end use of the fabric [37].This process includes diffusion of the dye into the liquid phase followed by adsorption onto the outer surface of the fibers, and finally diffusion and adsorption on the inner surface of the fibers [40]. Depending on the expected end use of the fabrics, different fastness properties may be required. For instance, swimsuits must not bleed in water and automotive fabrics should not fade after prolonged exposure to sunlight [37]. Different types of dye and chemical additives are used to obtain these properties, which is carried out during the finishing step. Dyeing can also be accomplished by applying pigments (pigments differ from dyes by not showing chemical or physical affinity for the fibers) together with binders (polymers which fix the pigment to the fibers) [39,41].

Finishing involves treatments with chemical compounds aimed at improving the quality of the fabric. Permanent press treatments, water proofing, softening, antistatic protection, soil resistance, stain release and microbial/fungal protection are all examples of fabric treatments applied in the finishing process [37].

Dyeing can be carried out as a continuous or batch process [37]. The most appropriate process to use depends on several factors, such as type of material (fiber, yarn, fabric, fabric construction, garment), generic type of fiber, size of dye batch and quality requirements for the dyed fabric, but batch processes are more commonly used to dye textile materials [10].

In continuous processing, heat and steam are applied to long rolls of fabric as they pass through a series of concentrated chemical solutions. The fabric retains the greater part of the chemicals while rinsing removes most of the preparation chemicals. Each time a fabric is passed through a solution, an amount of water equivalent to the weight of the fabric must be used [37].

In batch processing, sometimes called exhaust dyeing, since the dye is gradually transferred from the dye bath to the material being dyed over a relatively long period of time [10], the dyeing occurs in the presence of dilute chemicals in a closed equipment such as a kier, kettle, beam, jet or beck [37]. Unlike the continuous process, instead of being passed through various baths in a long series of equipment sections, in the batch process the fabric remains in a single piece of equipment, which is alternately filled with water and then drained, at each step of the process. Each time the fabric is exposed to a separate bath, it uses five to ten times its own weight in water [37].

Some batch dyeing machines only operate at temperatures up to 100ºC. However, the system can be pressurized, allowing for the use of temperatures above 100ºC. Cotton, rayon, nylon, wool and some other fibers dye well at temperatures of 100ºC or below. Polyester and some other synthetic fibers dye more easily at temperatures above 100ºC [10].

Since the degree of dye fixation depends on the nature of the fiber, it is important to consider this topic. The fibers used in the textile industry can be divided into two main groups denominated natural fibers and synthetic fibers [36,37]. Natural fibers are derived from the environment (plants or animals), such as wool, cotton, flax, silk, jute, hemp and sisal, most of which are based on cellulose and proteins. On the other hand, synthetic fibers are organic polymers, mostly derived from petroleum sources, for example, polyester, polyamide, rayon, acetate and acrylic [37,39,42]. The two most important textile fibers are cotton, the largest, and polyester [43,44].

Cotton has been used for over 7000 years, and consists of mainly cellulose, natural waxes and proteins. The large number of hydroxyl groups on the cellulose provides a great water absorption capacity [39].

Several aromatic polyesters have been synthesized and studied. Of these, polyethylene terephthalate (PET) and polybutylene terephthalate (PBT) have been produced commercially for more than 50 years. Amongst other uses, PET has been used worldwide for the production of synthetic fibers due to its good physical properties. PET is manufactured from ethylene glycol (EG) and terephthalic acid (TPA) or dimethyl terephthalate (DMT). The polymerization proceeds in two steps: esterification and condensation reactions [45]. The polymer produced after condensation is solidified by jets of cold water and cut into regular granules which frequently have a cubic form. Then, the polymer melt is spun and the yarns are solidified by a stream of cold air [39].

The dye can be fixed to the fiber by several mechanisms, generally in aqueous solution, and may involve primarily four types of interaction: ionic, Van der Waals and hydrogen interactions, and covalent bonds [5].

Ionic interactions result from interactions between oppositely charged ions present in the dyes and fibers, such as those between the positive center of the amino groups and carboxyl groups in the fiber and ionic charges on the dye molecule, and the ionic attraction between dye cations and anionic groups ($-SO_3^-$ and $-CO_2^-$) present in the acrylic fiber polymer molecules. Typical examples of this type of interaction can be found in the dyeing of wool, silk and polyamide [5,36,41].

Van der Waals interactions come from a close approach between the π orbitals of the dye molecule and the fiber, so that the dye molecules are firmly "anchored" to the fiber by an affinity process without forming an actual bond. Typical examples of this type of interaction are found in the dyeing of wool and polyester with dyes with a high affinity for cellulose [36].

Hydrogen interactions are formed between hydrogen atoms covalently bonded in the dye and free electron pairs of donor atoms in the center of the fiber. This interaction can be found in the dyeing of wool, silk and synthetic fibers such as ethyl cellulose [5,36].

Covalent bonds are formed between dye molecules containing reactive groups (electrophilic groups) and nucleophilic groups on the fiber, for example, the bond between a carbon atom of the reactive dye molecule and an oxygen, nitrogen or sulfur atom of a hydroxy, amino or thiol group present in the textile fiber. This type of bond can be found in the dyeing of cotton fiber [5,36,46].

2.1. Application Of Liposome-Based Technology In Textile Dyeing Process

There is increasing interest in the textile industry in the development of eco-friendly textile processing, in which the use of naturally occurring materials such as phospholipids, would become important [47]. Phospholipids are natural surfactants and in the presence of water, they organize themselves so as to reduce unfavorable interactions between their hydrophobic tails and the aqueous solution; their hydrophilic head groups exposed to the aqueous phase forming vesicles. Liposomes or phospholipid vesicles are featured by clearly separate hydrophilic and hydrophobic regions [34,48].

Liposomes were first produced in England in 1961 by Alec D. Bangham, who was studying phospholipids and blood clotting. He found that when phospholipids were added to water, they immediately formed a sphere, because one end of each molecule was water soluble, while the opposite end was water insoluble [49]. From a chemical point of view, the liposome is an amphoteric compound containing both positive and negative charges [34,50].

Liposomes are defined as a structure composed of lipid vesicle bilayers which can encapsulate hydrophobic or hydrophilic compounds in the lipid bilayer or in the aqueous volume, respectively [51]. These structures are usually made up of phosphatidylcholine (PC), which has a hydrophilic part consisting of phosphate and choline groups and a hydrophobic part composed of two hydrocarbon chains of variable length [49, 52,53].

Liposomes are often distinguished according to their number of lamellae and size. Small unilamellar vesicles (SUV), large unilamellar vesicles (LUV) and large multilamellar vesicles (MLV) or multivesicular vesicles (MVV) can be differentiated [49]. The diameters of liposomes vary from a nanometer to a micrometer [34]. Multilamellar liposomes (MLV) usually range from

500 to 10,000 nm. Unilamellar liposomes can be small (SUV) or large (LUV); SUV are usually smaller than 50 nm and LUV are usually larger than 50 nm. Very large liposomes are called giant liposomes (10,000 - 10,00,000 nm). They can be either unilamellar or multilamellar. The liposomes containing encapsulated vesicles are called multi-vesicular and they range from 2,000-40,000 nm. LUVs with an asymmetric distribution of phospholipids in the bilayers are called asymmetric liposomes [54]. The thickness of the membrane (phospholipid bilayer) measures approximately 5 to 6 nm [49].

According to Sivasankar, Katyayani (2011), the preparation of liposomes is based on lipids, and those normally used are:

Natural phospholipids:

- Phosphatidyl choline (PC) - Lecithin
- Phosphatidyl ethanolamine (PE) - Cephalin
- Phosphatidyl serine (PS)
- Phosphatidyl inositol (PI)
- Phosphatidyl glycerol (PG)

Synthetic phospholipids:

For saturated phospholipids:

- Dipalmitoyl phosphatidyl choline (DPPC)
- Distearoyl phosphatidyl choline (DSPC)
- Dipalmitoyl phosphatidyl ethanolamine (DPPE)
- Dipalmitoyl phosphatidyl serine (DPPS)
- Dipalmitoyl phosphatidic acid (DPPA)
- Dipalmitoyl phosphatidyl glycerol (DPPG)

For unsaturated phospholipids:

- Dioleoyl phosphatidyl choline (DOPC)
- Dioleoyl phosphatidyl glycerol (DOPG) [52].

Natural acidic lipids, such as PS, PG, PI, PA (phosphatidic acid) and cardiolipin (CL), are added when anionic liposomes are desired, and cholesterol is often included to stabilize the bilayer. These molecules are derivatives of glycerol with two alkyl groups and one amphoteric group [34].

Phosphatidylcholine is the biological lipid most widely used for producing liposomes. Liposomes based on phosphatidylcholine consist of phosphatidic acid and glycerin, with two alcohol groups esterified by fatty acids and a third group esterified by phosphoric acid, to which the amino alcohol choline is added as a polar group [34,49].

According to Barani, Montazer (2008), normally four different methods can be used for the preparation of liposomes:

1. Dry lipid film;
2. Emulsions;
3. Micelle-forming detergents;
4. Alcohol injection technology [34].

Liposomes have two distinct roles: they can provide an excellent model for biological membranes, and they are being developed as controlled delivery systems for hydrophilic and lipophilic agents [34,53,55]. They are promising candidates for adjuvant and carrier systems for drug delivery, are well-documented, and can be used for the same purpose in textile materials [33].

Encapsulation or liposome technology is applied in numerous fields, such as in pharmaceuticals, cosmetics, foods, detergents, textiles and other applications where it is important to liberate the encapsulated material slowly [34,50]. This new clean technology has already been adopted by some textile industries [52]. In recent years, liposomes have been examined as a way of delivering dyes to textiles in a cost-effective and environmentally sensitive way [56].

Conventional dyeing processes consume a great deal of energy, a significant amount of which is wasted in controlling the process parameters in order to achieve uniform results. With respect to the carrier role of liposomes, they can be used in several textile processes such as textile finishing and dyeing, with several types of dyes and fibers. They are nontoxic, biodegradable, and can encapsulate a wide range of solutes [34]. In addition, the main advantages of liposomes are a clear reduction in dyeing temperature (about 10°C as compared to conventional dyeing), improved quality of the textiles produced, with additional benefits with respect to material weight yield during subsequent spinning, improved smoothness and mechanical properties of the dyed textiles, and a clear reduction in the contamination load of the dye baths [52,57]. Low temperature gives a more natural feel and improved quality, with lower environmental impact [34].

In recent years, liposomes have been used in the textile industry as a carrier for auxiliary materials (leveling, retarding and wetting agents) in dyeing, mainly for wool dyeing, and for finishing processes [34,55]. One of the most common problems with textile auxiliaries is that they fail to form a complex in the solution bath. This problem can be solved by using liposomes with selected positive or negative charges. Liposomes can be prepared according to the type of process, solute material and fiber structure [34]. Liposomes from phospholipids have been widely used as a dye carrier in the dyeing process, and create eco-friendly textile processes. Due to their structural properties, liposomes can encapsulate hydrophilic dyes (reactive, acid and basic dyes) in the aqueous phase, and hydrophobic dyes (disperse dyes) in the phospholipid bilayers [58]. Liposomes containing a dye are generally large, irregular and unilamellar [50].

According to Barani & Montazer (2008), the application of liposomes in textile processing can be useful when the release of the solute material is important, and improves the final properties of the products. A wetting agent is required in the conventional bleaching bath of cotton fabrics, but this step can be eliminated by using liposomes. The presence of liposomes in the peroxide bleaching bath can improve the mechanical properties of fabrics and their brightness. Liposomes contain particles of oxidant present in the bleaching solution that represent an unusual reservoir, and release the bleaching agent gradually into the bleaching bath. Moreover, the encapsulation of catalysts used for the decomposition of hydrogen peroxide radicals can be another factor in retarding the rate of decomposition. In this way, liposomes act as a stabilizing agent in the bleaching bath [34].

The role of auxiliary products is very important in textile dyeing with disperse dyes [53]. These compounds show extremely low solubility in water and dispersing agents are needed to maintain a fine, stable dispersion throughout the whole dyeing process at the different temperatures. Martí et al. (2007) analyzed the usefulness of commercial textile liposomes as dispersing agents, and observed that liposomes could be considered as suitable dispersing auxiliaries for polyester dyeing at high temperatures, considering their capacity to stabilize dye dispersions and achieve a suitable dye exhaustion level, with the added value of their environmentally friendly nature [53]. Liposomes clearly improve the dispersion efficiency as compared to conventional dispersing agents [34].

Additionally, liposomes for textile use show a similar price to that of synthetic surfactants used in the dyeing of polyester with disperse dyes. However, the new technology is more environmentally friendly, and hence the reduction in the environmental problem can lead to economic advantages [53]. In addition, liposome preparations tend not to foam. This is an advantage that distinguishes liposomes from other textile auxiliaries [34].

According to Martí et al. (2010), the dyeing of wool and wool blends with the aid of liposomes has demonstrated better quality, energy saving and a reduction in the environmental impact and also the temperature could be reduced, resulting in less fiber damage. Moreover, dye bath exhaustion was shown to be over 90% at the lower temperature (80°C) used, resulting in significant savings in energy costs [55]. The impact of the dyeing process on the environment was also considerably lower, the COD being reduced by about 1000 units [53].

Therefore, liposome-based technology is an alternative, eco-friendly method, which could reduce the environmental impact, offering technical and economic advantages for the textile industry.

2.2. Effect Of Ultrasonic Energy On The Dyeing Process

Ultrasound-assisted textile dyeing was first reported by Sokolov and Tumansky in 1941[59]. The basic idea of this technology is that ultrasound can enhance mass transfer by reducing the stagnant cores in the yarns. The improvements observed are generally attributed to cavitation phenomena and to other resulting physical effects such as dye dispersion (breaking up of aggregates with high relative molecular mass), degassing (expulsion of dissolved or entrapped air from the fiber capillaries), strong agitation of the liquid (reduction in thickness of the fiber-liquid boundary layer), and swelling (enhancement of dye diffusion rate inside the fiber) [59,60].

According to Vankar & Shanker (2008), ultrasound allows for process acceleration, obtaining the same or better results than existing techniques, but under less extreme conditions, i.e., lower temperatures and lower concentrations of the chemicals used. Wet textile processes assisted by ultrasound are of great interest to the textile industry for this reason [61], and Khatri et al. (2011) showed that the dyeing of polyester fiber using ultrasonic energy resulted in an increased dye uptake and enhanced dyeing rate [35].

Due to the revolution in environmental protection, the use of ultrasonic energy as a renewable source of energy in textile dyeing has been increased, due to the variety of advantages associated with it. On the other hand, there is a growing demand for natural, eco-friendly dyeing for the health sensitive application to textile garments as an alternative to harmful synthetic dyes, which poses a need for suitable effective dyeing methodologies [62].

Ultrasonic energy can clean or homogenize materials, accelerating both physical and chemical reactions, and these qualities can be used to improve textile processing methods. Environmental concern has been focused on textile processing methods for quite some time, and the use of ultrasonic energy has been widely studied in terms of improving washing fastness. The textile dyeing industry has long been struggling to cope with high energy costs, rapid technological changes and the need for a faster delivery time, and the effective management of ultrasonic energy could reduce energy costs and improve productivity [35]. Ultrasonic waves are vibrations with frequencies above 17 kHz, out of the audible range for humans, requiring a medium with elastic properties for propagation. The formation and collapse of the bubbles formed by ultrasonic waves (known as cavitation) is generally considered to be responsible for most of the physical and chemical effects of ultrasound in solid/liquid or liquid/liquid systems [63]. Cavitation is the formation of gas-filled microbubbles or cavities in a liquid, their growth, and under proper conditions, their implosive collapse [59].

It has been reported that ultrasonic energy can be applied successfully to wet textile processes, for example laundering, desizing, scouring, bleaching, mercerization of cotton fabrics, enzymatic treatment, dyeing and leather processing, together with the decoloration/mineralization of textile dyes in waste water [60].

In addition, ultrasonic irradiation shows promise, and has the potential, for use in environmental remediation, due to the formation of highly concentrated oxidizing species such as hydroxyl radicals (HO•), hydrogen radicals (H•), hydroperoxyl radicals (HO_2•) and H_2O_2, and localized high temperatures and pressures [59]. Therefore, the use of ultrasonic energy could indeed reduce the environmental impact caused by the textile industry.

3. FINISHING AND WASTE WATER

The contamination of natural waters has become one of the biggest problems in modern society, and the economical use of this natural resource in production processes has gained special attention, since in predictions for the coming years, the amount of water required per capita is of concern. This environmental problem is related not only to its waste through misuse, but also to the release of industrial and domestic effluents [64].

Of the industries with high-polluting power, the textile dyeing industry, responsible for dyeing various types of fiber, stands out. Independent of the characteristics of the dyes chosen, the final operation of all dyeing process involves washing in baths to remove excesses of the original or hydrolyzed dyes not fixed to the fiber in the previous steps [36]. In these baths, as previously mentioned, it is estimated that approximately 10-50% of the dyes used in the dyeing process are lost, and end up in the effluent [17,21,22], contaminating the environment with about one million tons of these compounds [65]. The dyes end up in the water bodies due mainly to the use of the activated sludge treatment in the effluent treatment plants, which has been shown to be ineffective in removing the toxicity and coloring of some types of dye [33,60,66,67]. Moreover, the reduction of azo dyes by sodium hydrosulfite and the successive chlorination steps with hypochlorous acid, can form 2-benzotriazoles fenilbenzotriazol (PBTA) derivatives and highly mutagenic aromatic amines, often more mutagenic than the original dye [68]. In an aquatic environment, this dye reduction can occur in two phases: 1) The application of reducing agents to the newly-dyed fibers to remove the excess unbound dye, which could lead to "bleeding" of the fabrics during washing, and 2) The use of reducing agents in the bleaching process, in order to make the effluent colorless and conform with the legislation. This reduced colorless effluent containing dyes is sent to the municipal sewage treatment plant, where they chlorinate the effluents before releasing them into water bodies where they may generate PBTAs. Several different PBTAs are already described in the literature, and their chemical structures vary depending on the dyes that originated them [63,69].

So the release of improperly treated textile effluents into the environment can become an important source of problems for human and environmental health. The major source of dye loss corresponds to the incomplete fixation of the dyes during the textile fiber dyeing step [36].

In addition to the problem caused by the loss of dye during the dyeing process, within the context of environmental pollution, the textile industry is also focused due to the large volumes of water used by its industrial park, consequently generating large volumes of effluent [64]. It has been calculated that approximately 200 liters of water are needed for each kilogram of cotton produced [70]. These effluents are complex mixtures of many pollutants, ranging from original colors lost during the dyeing process, to associated pesticides and heavy metals [71], and when not properly treated, can cause serious contamination of the water sources [64]. So the materials that end up in the water bodies are effluents containing a high organic load and biochemical oxygen demand, low dissolved oxygen concentrations, strong color and low biodegradability. In addition to visual pollution, the pollution of water bodies with these compounds causes changes in the biological cycles of the aquatic biota, particularly affecting the photosynthesis and oxygenation processes of the water body, for example by hindering the passage of sunlight through the water [72].

Moreover, studies have shown that some classes of dye, especially azo dyes and their by-products, may be carcinogenic and / or mutagenic [27,33,67,73-77], endangering human health, since the wastewater treatment systems and water treatment plants (WTP) are ineffective in removing the color and the mutagenic properties of some dyes [78,79]. The difficulty in removing them from the environment can be attributed to the high stability of these compounds, since they are designed to resist biodegradation to meet the demands of the consumer market with respect to durability of the colors in the fibers, consequently implying that they also remain in the environment for a long time [32].

With respect to the legislation, there is no consensus amongst the different countries concerning effluent discharge, and there is no official document listing the different effluent limit values applied in different countries. Many federal countries, such as the United States of America, Canada and Australia have national environmental legislation, which, as in Europe, establishes the limits that must be complied with. Some countries, such as Thailand, have copied the American system, whereas others, such as Turkey or Morocco, have copied the European model. In some countries, for example India, Pakistan and Malaysia, the emission limits are recommended, but are not mandatory [80]. With respect to the color, in some countries such as France, Austria and Italy, there are limits for the color of the effluent, but since they use different units, a comparison is impossible. The oldest unit is the Hazen, in use since the beginning of the 20th century, but in France, the current unit is (mg L^{-1}Pt–Co). The coloration values are determined by a comparative analysis with model solutions prepared according to defined procedures [80].

Based on all the problems cited above regarding the discharge of effluents into the environment, it is obvious there is a need to find alternative treatments that are effective in removing dyes from effluents.

4. WASTE WATER TREATMENT

These days, environmental pollution can undoubtedly be regarded as one of the main problems in developed and developing countries. This is due, not just to one, but to a number of factors, such as the misuse of natural resources, inefficient legislation and a lack of environmental awareness. Fortunately, in recent years there has been a trend for change and a series of scientific studies are being used as an important tool in the development of new treatment technologies and even in the implementation of processes and environmentally friendly actions [64,81-84].

Every industrial process is characterized by the use of inputs (raw materials, water, energy, etc.) that undergo transformation giving rise to products, byproducts and waste. The wastes produced at all stages of the various types of human activity, both in terms of composition and volume, vary according to the consumption practices and production methods. The main concerns are focused on the impact these can have on human health and the environment. Hazardous waste, produced mainly by industry, is particularly worrying, because when incorrectly managed, it becomes a serious threat to the environment and therefore to human health. Thus the study of new alternatives for the treatment of different types of industrial effluent continues to be a challenge to combat anthropogenic contamination.

Amongst that of several other industries, the textile sector waste has received considerable attention in recent years, since it can generate large volumes of effluents that, if not correctly treated before being disposed into water resources, can be a problem, as previously mentioned. Effluents from the textile industry are extremely complex, since they contain a large variety of dyes, additives and derivatives that change seasonally, increasing the challenge to find effective, feasible treatments. Currently, the processes developed and available for these industries are based on methods that were designed for other waste, and have limitations when applied to textile effluents. As a consequence, these industries produce colored wastewater with a high organic load, which can contribute enormously to the environmental pollution of surface water and treatment plants if not properly treated before disposal into the water resources [85]. The ingestion of water contaminated with textile dyes can cause serious damage to the health of humans and of other living organisms, due to the toxicity, highlighting mutagenicity of its components [86,87]. Therefore treatments that are more efficient and economical than those currently available are required.

There are several techniques for the treatment of effluents, such as incineration, biological treatment, absorption onto solid matrices, etc. However, these techniques have their drawbacks, such as the formation of dioxins and furans, caused by incomplete combustion during incineration; long periods for biological treatment to have an effect, as also the adsorptive process, that is based on the phase transfer of contaminants without actually destroying them [88,89]. The problem is further aggravated in the textile industry effluents, due to the complexity of their make-up. Thus it can be seen that processes are being used that are not entirely appropriate for the treatment of textile effluents, thereby creating a major challenge for the industry and laundries that need to adapt to current regulations for the control of the color of effluents with a high organic load.

The use of filtration membranes and/or separation [90] and biological methods [91], in addition to incineration processes involving adsorption onto solid matrices, has also being adopted by the textile industry and is receiving considerable attention. However, all these processes only involve phase transfer, generating large amounts of sludge deposited at the end of the tanks and low efficiency in color removal and reduction of the organic load. According to this scenario, many studies have been carried out with the aim of developing new technologies capable of minimizing the volume and toxicity of industrial effluents. Unfortunately, the applicability of these types of system is subject to the development of modified procedures and the establishment of effluent recycling systems, activities that imply evolutionary technologies and which are not yet universally available. Thus the study of new alternatives for the treatment of many industrial effluents currently produced is still one of the main weapons to combat the phenomenon of anthropogenic contamination.

Due to their considerable danger, several authors have attempted to find new forms of treatment to reduce the serious environmental and toxicological risks caused by various organic compounds. Amongst the many reported cases are those based on the use of specific microorganisms, and degradation using advanced oxidation processes (AOP) such as Fenton, photo-Fenton and heterogeneous photocatalysis, which are highlighted below.

The use of microorganisms cultivated specifically for the degradation of pollutents to increase the yield of degradation, has been reported by some authors. For example, Flores et al. (1997) examined the behavior of 25 N-substituted aromatic compounds such as organic compounds,azo dyes and nitro, using the methanogenic bacteria acetoclastic, and found that under anaerobic conditions it was easy to mineralize various of the compounds evaluated with a good yield, especially the nitroaromatic and azo dyes [91].

Bornick et al. (2001) evaluated the use of aerobic microorganisms to degrade aromatic amines present in sediments of the river Elbe in Germany. The results obtained showed that it was possible to predict the qualitative degradation of the aromatic amines using degradation constants [92].

Using a structural design, Wang et al.(2007) isolated a bacterium capable of promoting the degradation of the compounds pentyl amine and aniline present in water oil extraction in China. Under conditions of neutral pH and complete aeration of 6 mg O_2 / l at a temperature of 30 ° C, they obtained degradation yields of 82% and 78%, respectively, for pentyl amine and aniline [93].

However, in general, although the use of microorganisms in the treatment of industrial and laboratory wastes containing aromatic amines deserves attention, mainly due to the low investment and maintenance costs, the results are far from ideal, due to the low biodegradation yields, long treatment times, and the generation of sludge deposited at the bottom of the treatment ponds [94].

Of the studies carried out using a source of hydroxyl radicals with oxidizing agents, the photo Fenton system developed by Fukushima et al., 2000 to promote the degradation of aniline stands out. This method has shown promise for the mineralization of aromatic amines, obtaining a reduction of approximately 85%. However, high performance liquid chromatography (HPLC) identified a number of intermediate species formed during the degradation of the aniline, such as p-aminophenol, p-hydroquinone, maleic and fumaric acids and NH_4^+ [95].

Studies involving heterogeneous photocatalysis also deserve attention: for example Pramauro et al., (1995) promoted the degradation of various aniline derivatives using TiO_2 particles suspended in a solution. Under optimal conditions, the method developed showed rapid mineralization of the aromatic amines examined in less than 1 hour of analysis. By-products generated during the degradation of these compounds were identified at the start of the reaction, but none were identified at the end of the reaction. The use of solar radiation was also evaluated, but was shown to be less efficient than artificial radiation [96].

Augugliaro et al.(2000), confirmed that heterogeneous photocatalysis using TiO_2 as a semiconductor may be a suitable method for the complete photodegradation of aniline, 4-ethylaniline and 4-chloroaniline in an aqueous medium. The kinetic parameters for the Langmuir-Hinshelwood model were used to describe the importance of the adsorption results, which proved to be independent of the pH of the solution and of the type of substituent on the aromatic ring of the amine [97].

Canle et al. (2005) studied the behavior of the adsorption of aniline and dimethyl aniline onto three species of TiO_2 (P25, anatase and rutile) and established that the greatest adsorption occurred onto TiO_2 P25. The effect of pH on the degradation provided by these compounds was also evaluated, and lower mineralization percentages were observed in acidic media. This type of behavior can be attributed to the positive charges on both the aniline and the semiconductor, providing electrostatic repulsion between the two species. Thus, an alkaline medium has been recommended as the most appropriate one to promote the mineralization of aromatic amines using a TiO_2 P25 type semiconductor [98].

Chu et al. (2007) observed the effects of pH variation and the addition of hydrogen peroxide on the degradation of 2-chloroaniline using TiO_2 as a semiconductor, with and without the application of UV radiation. The results showed that the addition of low concentrations of H_2O_2 to the UV/TiO_2system provided a significant increase in degradation of the aromatic amine. The addition of an excess of H_2O_2 promoted no increase in degradation, as expected; to the contrary, a reduction in the reaction rate was observed. The variation in pH was evaluated in both systems, and the condition leading to the highest percentages of mineralization was obtained in an alkaline medium using H_2O_2/UV/TiO_2 [99].

Low et al. (1991) monitored the inorganic products resulting from the degradation of several organic nitrogenated, sulfured and halogenated compounds. Degradation was carried out using TiO_2 as the semiconductor and artificially illuminated UV radiation. Ammonium ions were found to be present in higher concentrations than nitrate ions, which can be explained by the fact that compounds having a nitrogen element in their structure pass through a complex degradation step where the generation of ammonium ions is more favorable than the generation of nitrate ions. In turn, compounds with nitro groups in their structures, showed higher concentrations of nitrate ions. Under ideal conditions, all the elements were converted into their respective inorganic forms. Organic carbon was converted to CO_2, the halogens to their corresponding halide, sulfur compounds to sulfate, the phosphate to phosphorus and nitrogen to ammonium and nitrate [100]. However, all these studies used a photocatalytic titanium suspension, requiring a subsequent step to remove the semiconductor.

With a view to these problems, a new technique that has been studied recently, with significant success, is the oxidation of organic matter via the generation of hydroxyl radicals (OH•). This kind of effluent treatment has been highlighted for its destructive character with respect to the organic matter. The advanced oxidation processes (AOP) are processes based on the generation of highly oxidizing species, the hydroxyl radical (OH•), which can oxidize the organic contaminants present in water, air or soil. These radicals have a high oxidation potential (E * = +2.72 V vs. the normal hydrogen electrode, NHE), which results in high reactivity with organic pollutants. This may initiate different types of reaction with different functional groups in organic compounds, forming unstable organic radicals which are then easily oxidized to CO_2, H_2O and inorganic acids, derived from this heteroatom. Many techniques have being developed using this principle as the method of treatment, and have shown potential success in organic residue mineralization processes [92,96, 101-103].

The association of the electrochemical properties with the photocatalytic properties of a semiconductor, have allowed for the development of a promising system, representing a very efficient technique as an AOP method that has received much attention in recent years, assigned as a photoelectrochemical process. The process gained notoriety for the possibility of forming hydroxyl radicals via the oxidation of water. The technique is based on the action of ultraviolet light (hυ) on a semiconductor capable of generating charges and e^-/ H^+, whose separation is facilitated by applying a positive potential (E_{APP}), greater than the potential of a flat band photocatalytic material. The generation of a potential gradient in the photoactivated semiconductor directs the electrons to an auxiliary electrode (cathode), delaying recombination between the holes (h^+) generated in the valence band (VB), and making them available for oxidation processes and the generation of hydroxyl radicals of interest [104].

The results of this process were very promising because of the relatively short treatment time but with great efficiency, both in the removal of color and in the reduction of the organic load. However, the limitations of this technique are related mainly to the choice of the ideal catalyst for promoting the generation of these oxidizing species. Catalysts that promote the generation of radicals absorbing radiation in the visible spectral region are the most desirable for this type of reaction, due to the large percentage emitted in the solar spectrum (approximately 45%) [104].

Thus, the development of an ideal process that promotes color removal and a reduction in the organic load of wastewater from the textile industry with great efficiency is a major challenge in all fields of science, since the synthesis of the best catalyst to take advantage of solar radiation, thus reducing the operating costs, and at the same time solve the problems involved in the hydrodynamics of the reactors, is of importance in the development of the treatment.

The expectations for developing an effective method for the treatment of these wastes are quite promising, but require continuous optimization and knowledge of new aspects. These include better fixation of the dyes to the fibers, process with less water consumption, less hazardous dyes with respect to human health and methods capable of identifying these compounds with more efficacy and rapidity and assays to identify any potential carcinogenic and / or mutagenic properties in the dyes and their derivatives; genetic improvements to produce more efficient culture mediums and resistant biological treatments, leading to a reduction in the generation of sludge; the synthesis of materials that catalyze reactions in the visible spectral regions, leading to a more economic photoeletrochemical method, and also new engineering advances for the construction of more effective reactors, which can take advantage of all these developments in an integrated system, extending the performance of a process more appropriate for the treatment of such a complex effluent.

5. OPTIMIZATION OF THE DYEING PROCESSES TO REDUCE THE ENVIRONMENTAL IMPACT OF THE TEXTILE INDUSTRY

The search and development of new methods to promote the treatment of effluents from the textile industry with a maximum of efficiency of the process of decolorization and / or removal of these compounds present in the medium can trigger further damage human health and the environment is fundamental importance. The understanding of the composition of waste generated is extremely significant to develop these methods of treatment due to the high complexity by virtue of huge number of compounds which are added at different stages of the dyeing fabrics.

Environmental problems with used dye baths are related to the wide variety of different components added to the dye bath, often in relatively high concentrations. In the future, many of textile factories will face the requirement of reusing a significant part of all incoming freshwater because traditionally used methods are insufficient for obtaining the required water quality.

However, due to dwindling supply and increasing demand of water in the textile industries, a better alternative is to attempt to further elevate the water quality of wastewater effluent from a secondary wastewater treatment plant to a higher standard for reuse. Thus far very little attention has been paid to this aspect [105].

Therefore, the investment in the search for methodologies to more effective treatment of these effluents can be much smaller than that spent in tertiary treatment to remove these products in low level of concentrations and in the presence of much other interference. This requires action that the cost / benefit are reviewed and the development of new techniques for wastewater treatment capable of effective removal of these dyes is intensified and made economically viable [105,106].

An alternative to minimize the problems related to the treatment of textile effluents would be the development of more effective dye that can be fixed fiber with higher efficiency decreasing losses on tailings waters and reducing the amount of dye required in the dyeing process, reducing certainly improve the cost and quality of the effluent.

6. CONCLUSION

It was concluded that the synthetic textile dyes represent a large group of organic compounds that could have undesirable effects on the environment, and in addition, some of them can pose risks to humans. The increasing complexity and difficulty in treating textile wastes has led to a constant search for new methods that are effective and economically viable. However, up to the present moment, no efficient method capable of removing both the color and the toxic properties of the dyes released into the environment has been found.

ACKNOWLEDGMENTS

This work was supported by the Faculty of Pharmaceutical Sciences at Ribeirão Preto - University of São Paulo, Brazil, and by FAPESP, CAPES and CNPq.

REFERENCES

1. Kirk-Othmer. Encyclopedia of Chemical Technology, v. 7, 5th Edition. Wiley-Interscience; 2004.

2. Bafana A, Devi SS, Chakrabarti T. Azo dyes: past, present and the future. Environmental Reviews 2011; 19 350–370.

3. Zollinger, H. Synthesis, Properties of Organic Dyes and Pigments. In: Color Chemistry. New York, USA: VCH Publishers; 1987. p. 92-102.

4. Carneiro PA, Nogueira, RFP, Zanoni, MVB. Homogeneous photodegradation of C.I. Reactive Blue 4 using a photo-Fenton process under artificial and solar irradiation. Dyes and Pigments 2007; 74 127-132.

5. Christie R. Colour Chemistry. Cambridge, United Kingdom: The Royal Society of Chemistry; 2001.

6. Dos Santos AB, Cervantes FJ, van Lier JB. Review paper on current technologies for decolourisation of textile wastewaters: Perspectives for anaerobic biotechnology. Bioresource Technology 2007; 98 (12) 2369-2385.

7. Arun Prasad AS, Bhaskara Rao KV. Physico chemical characterization of textile effluent and screening for dye decolorizing bactéria. Global Journal of Biotechnology and Biochemistry 2010; 5(2) 80-86.

8. Robinson T, McMullan G, Marchant R, Nigam P. Remediation of dyes in textile effluent: a critical review on current treatment technologies with a proposed alternative. Bioresource Technology 2001; 77 (12) 247-255.

9. Ogugbue CJ, Sawidis T. Bioremediation and Detoxification of Synthetic Wastewater Containing Triarylmethane Dyes by Aeromonas hydrophila Isolated from Industrial Effluent. Biotechnology Research International 2011; DOI 10.4061/2011/967925.

10. Perkins WS. A Review of Textile Dyeing Processes. American Association of Textile Chemists and Colorists 1991; 23 (8): 23–27.http://www.revistavirtualpro.com/files/TIE08_200704.pdf (accessed 12 May 2012)

11. Couto SR. Dye removal by immobilised fungi. Biotechnology Advances 2009; 27(3) 227-235.

12. O'Neill C, Hawkes FR, Hawkes DL, Lourenço ND, Pinheiro HM, Delée W. Colour in textile effluents – sources, measurement, discharge consents

and simulation: a review. Journal of Chemical Technology and Biotechnology 1999; 74 (11) 1009-1018.

13. Forgacs E, Cserháti T, Oros G. Removal of synthetic dyes from wastewaters: a review. Environment International 2004; 30 (7) 953-971

14. Przystaś W, Zabłocka-Godlewska E, Grabińska-Sota E. Biological Removal of Azo and Triphenylmethane Dyes and Toxicity of Process By-Products. Water Air Soil Pollut 2012; 223 (4) 1581-1592.

15. Sen S, Demirer GN. Anaerobic treatment of real textile wastewater with a fluidized bed reactor. Water Research 2003; 37 (8) 1868-1878.

16. Dos Santos, A B. Reductive Decolourisation of Dyes by Thermophilic Anaerobic Granular Sludge. PhD-Thesis. Wageningen University, The Netherlands, 2005.

17. Ben Mansour H, Houas I, Montassar F, Ghedira K, Barillier D, Mosrati R, Chekir-Ghedira L. Alteration of in vitro and acute in vivo toxicity of textile dyeing wastewater after chemical and biological remediation. Environmental science and pollution research international 2012; DOI 10.1007/s11356-012-0802-7.

18. Talarposhti AM, Donnelly T, Anderson GK. Colour removal from a simulated dye wastewater using a two-phase anaerobic packed bed reactor. Water Research 2001; 35 (2) 425-432.

19. Ibrahim MB, Poonam N, Datel S, Roger M. Microbial decolorization of textile dye-containing effluents: a review, Bioresource Technology 1996; 58(3) 217-227.

20. Wijetunga S, Li XF, Jian C. Effect of organic load on decolourization of textile wastewater containing acid dyes in upflow anaerobic sludge blanket reactor. Journal of Hazardous Materials 2010; 177 (1-3) 792-798.

21. Vaidya AA, Datye KV. Environmental pollution during chemical processing of synthetic fibers. Colourage 1982; 14 3-10.

22. Rajaguru P, Fairbairn LJ, Ashby J, Willington MA, Turner S, Woolford LA, Chinnasamy N, Rafferty JA. Genotoxicity studies on the azo dye Direct

Red 2 using the in vivo mouse bone marrow micronucleus test. Mutation Research 1999; 444(1) 175-180.

23. Hubbe MA, Beck KR, O'Neal WG, Sharma YC. Cellulosic substrates for removal of pollutants from aqueous systems: a review. 2. Dyes. Dye biosorption: Review. BioResources 2012; 7 (2), 2592-2687.

24. Carliell CM, Barclay SJ, Shaw C, Wheatley AD, Buckley C A. The effect of salts used in textile dyeing on microbial decolourisation of a reactive azo dye. Environmental Technology 1998; 19 (11) 1133-1137.

25. Seesuriyachan P, Takenaka S, Kuntiya A, Klayraung S, Murakami S, Aoki K. Metabolism of azo dyes by Lactobacillus casei TISTR 1500 and effects of various factors on decolorization. Water Research 2007; 41(5) 985-992.

26. Ben Mansour H, Corroler D, Barillier D, Ghedira K, Chekir L, Mosrati R. Evaluation of genotoxicity and pro-oxidant effect of the azo dyes: Acids yellow 17, violet 7 and orange 52, and of their degradation products by Pseudomonas putida mt-2. Food and Chemical Toxicology 2007; 45 (9) 1670-1677.

27. Chung KT, Cerniglia CE. Mutagenicity of azo dyes: Structure-activity relationships. Mutation Research, 1992; 277 (3) 201-220.

28. Pinheiro HM, Touraud E, Thomas O. Aromatic amines from azo dye reduction: status review with emphasis on direct UV spectrophotometric detection in textile industry wastewaters. Dyes and Pigments 2004; 61 (2) 121-139.

29. Umbuzeiro GA, Freeman H, Warren SH, Kummrow F, Claxton LD. Mutagenicity evaluation of the commercial product CI Disperse Blue 291 using different protocols of the Salmonella assay. Food and Chemical Toxicology 2005; 43 (1) 49-56.

30. Arlt VM, Glatt H, Muckel E, Pabel U, Sorg BL, Schmeiser HH, Phillips DH. Metabolic activation of the environmental contaminant 3-nitrobenzanthrone by human acetyltransferases and sulfotransferase. Carcinogenesis 2002; 23 (11) 1937-1945.

31. Hao OJ, Kim H, Chiang PC, 2000. Decolorization of wastewater. Critical Reviews in Environmental Science and Technology 2000; 30 (4) 449-505.

32. Firmino PIM, Silva MER, Cervantes FJ, Santos AB. Colour removal of dyes from synthetic and real textile wastewaters in one- and two-stage anaerobic systems. Bioresource Technology 2010; 101(20) 7773-7779.

33. Carneiro PA, Umbuzeiro GA, Oliveira DP, Zanoni MVB. Assessment of water contamination caused by a mutagenic textile effluent/dyehouse effluent bearing disperse dyes. Journal of Hazardous Materials 2010; 174 (1-3) 694-699.

34. Barani H, Montazer M. A Review on Applications of Liposomes in Textile Processing. Journal of Liposome Research 2008; 18 (3) 249-262.

35. Khatri Z, Memon MH, Khatri A, Tanwari A. Cold Pad-Batch dyeing method for cotton fabric dyeing with reactive dyes using ultrasonic energy. Ultrasonics Sonochemistry 2011 18 (6) 1301-1307.

36. Guaratini CCI, Zanoni MVB. Textile dyes. Química Nova 2000;23(1): 71-78. http://www.scielo.br/pdf/qn/v23n1/2146.pdf (accessed 01 May 2012).

37. Moore SB, Ausley LW. Systems thinking and green chemistry in the textile industry: concepts, technologies and benefits. Journal of Cleaner Production 2004;12 585–601. DOI: 10.1016/S0959-6526(03)00058-1

38. Reddy SS, Kotaiah B, Reddy NSP. Color pollution control in textile dyeing industry effluents using tannery sludge derived activated carbon. Bulletin of the Chemical Society of Ethiopia 2008;22 (3): 369-378. http://www.ajol.info/index.php/bcse/article/viewFile/ 61211/ 49389 (accessed 16 May 2012).

39. Alcantara MR, Daltin D. A química do processamento têxtil. Química Nova 1996;19(3): 320-330 http://www.quimicanova.sbq.org.br/qn/qnol/1996/vol19n3/v19_n3_17.pdf (accessed 19 April 2012).

40. Vassileva V, Valcheva E, Zheleva Z. The kinetic model of reactive dye fixation on cotton fibers. Journal of the University of Chemical Technology and Metallurgy 2008;43 (3): 323-326. http:// www.uctm.edu/j2008-3/8_V_Vasileva_323-326.pdf (accessed 10 May 2012).

41. Zollinger H. Color Chemistry. Syntheses,properties and applications of organic dyes and pigments. Weinheim: VCH Publishers; 2003.

42. Candlin J. Polymers. Dyestuffs in the Chemical Industry. London: Chapman & Hall; 1994.

43. Gregory P. Dyestuffs. In: Heaton C.A. (ed.) The Chemical Industry. London: Chapman & Hall; 1994. p143-188.

44. Gregory P. 2009. Dyes and Dye Intermediates. In: Kirk-Othmer (ed.) Encyclopedia of Chemical Technology. New Jersey: John Wiley & Sons; 2009. p1–66. DOI: 10.1002/0471238961.0425051907180507.a01.pub2

45. Pang K, Kotek R, Tonelli A. Review of conventional and novel polymerization processes for polyesters. Progress in Polymer Science 2006;31(11) 1009–1037. DOI: 10.1016/j.progpolymsci.2006.08.008

46. Hunger K., editor. Industrial Dyes: Chemistry, Properties, Applications. Weinheim: Wiley-VCH Verlag GmbH & Co. KGaA; 2004. DOI: 10.1002/3527602011

47. Baptista ALF, Coutinho PJG,: Real Oliveira MECD, Rocha Gomes JIN. Effect of surfactants in soybean Lecithin liposomes studied by energy transfer between NBD-PE and N-Rh-PE. Journal of Liposome Research 2000; 10 (4) 419-429.

48. Lesoin L, Crampon C, Boutin O, Badens E. Preparation of liposomes using the supercritical anti-solvent (SAS) process and comparison with a conventional method. The Journal of Supercritical Fluids 2011; 57 162-174

49. Sivasankar, M, Katyayani, T. Liposomes – The future of formulations. International Journal of Research in Pharmacy and Chemistry 2011; 1(2) 259-267.

50. El-Zawahry MM, El-Shami S, El-Mallah S.H. Optimizing a wool dyeing process with reactive dye by liposome microencapsulation. Dyes and Pigments 2007; 74 684-691.

51. Martí M, de la Maza A, Parra JL, Coderch L. Liposome as dispersing agent into disperse dye formulation. Textile Research Journal 2011; 81(4) 379-387.

52. Montazer M., Validi M., Toliyat T. Influence of Temperature on Stability of Multilamellar Liposomes in Wool Dyeing. Journal of Liposome Research, 2006; 16:81-89.

53. Martí M, Coderch L, de la Maza A, Parra JL. Liposomes of phosphatidylcholine: a biological natural surfactant as a dispersing agent. Color Technology 2007; 123 237-241.

54. Riaz M. Liposomes preparation methods. Pakistan Journal of Pharmaceutical Sciences, 1996; 19(1) 65-77.

55. Martí M, de la Maza A, Parra JL, Coderch L. Liposome as dispersing agent into disperse dye formulation. Textile Research Journal 2010; 81(4) 379–387.

56. Nelson, G. Application of microencapsulation in textile. International Journal of Pharmaceutics 2002; 242 55-62.

57. Martí M, de la Maza A, Parra JL, Coderch L. Dyeing wool at low temperatures: new method using liposomes. Textile Research Journal 2001; 71(8) 678-682.

58. El-Zawahry MM, El-Mallahb MH, El-Shamib S. An innovative study on dyeing silk fabrics by modified phospholipid liposomes. Coloration Technology 2009; 125 164-171.

59. Vajnhandl S, Le Marechal AM. Ultrasound in textile dyeing and the decolouration/mineralization of textile dyes. Dyes and Pigments 2005; 65 89-101.

60. Ferrero F., Periolatto M. Ultrasound for low temperature dyeing of wool with acid dye. Ultrasonics Sonochemistry 2012; 19 601-606.

61. Vankar PS, Shanker R. Ecofriendly ultrasonic natural dyeing of cotton fabric with enzyme pretreatments. Desalination 2008; 230 62-69.

62. Mansour HF, Heffernan S. Environmental aspects on dyeing silk fabric with sticta coronate lichen using ultrasonic energy and mild mordants. Clean Technologies Environmental Policy 2011; 13 207-213.

63. Abou-Okeil A, El-Shafie A, El Zawahry MM. Ecofriendly laccase–hydrogen peroxide/ultrasound-assisted bleaching of linen fabrics and its influence on dyeing efficiency. Ultrasonics Sonochemistry 2010; 17 383-390.

64. Kunz A, Zamora PP, Moraes SG, Durán N. New tendencies on the textile effluent treatment. Quimica Nova 2002;25(1): 78-82. http://www.

scielo.br/scielo.php?script=sci_arttext&pid=S0100-40422002000 100014 (accessed 01 May 2012).

65. Associação Brasileira de Química. ABQ. Processos oxidativos avançados e tratamento de corantes sintéticos. Anais da ABQ. http:// www.unb.br/resqui/abq2004-1.pdf (accessed 18 May 2012).

66. Umbuzeiro GA, Roubicek DA, Rech CM, Sato MIZ, CLAXTON LD. Investigating the sources of the mutagenic activity found in a river using the Salmonella assay and different water extraction procedures. Chemosphere 2004; 54(11) 1589-1597. DOI: 10.1016/j. chemosphere. 2003.09.009

67. Umbuzeiro GA, Freeman HS, Warren SH, Oliveira DP, Terao Y, Watanabe T, Claxton LD. The contribution of azo dyes to the mutagenic activity of the Cristais river. Chemosphere 2005; 60 (1) 55-64. DOI:10.1016/j.chemosphere.2004.11.100

68. Shiozawa T, Suyama K, Nakano K, Nukaya H, Sawanishi H, Oguri A, Wakabayashi K, Terao Y. Mutagenic activity of 2 phenylbenzotriazole derivatives related to a mutagen, PBTA-1, in river water. Mutation Research 1999; 442 (2) 105-111.

69. Oliveira DP. Corantes como importante classe de contaminantes ambientais – um estudo de caso. PhD thesis. Universidade de São Paulo; 2005.

70. Carneiro PA, Osugi ME, Sene JJ, Anderson MA, Zanoni MVB. Evaluation of color removal and degradation of a reactive textile azo dye on nanoporous TiO thin-film electrodes. Electrochimica Acta 2004; 49 (22-23) 3807–3820. DOI: 10.1016/j.electacta.2003.12.057

71. McMullan G, Meehan C, Conneely A, Kirby N, Robinson T, Nigam P, Banat IM, Marchant R, Smyth WF. Mini-review: microbial decolorisation and degradation of textile dyes. Applied Microbiology and Biotechnology 2001; 56 (1-2): 81-87. DOI: 10.1007/ s002530000587. http://www.springerlink.com/content/ 04vuqljha jpa32rv/ (accessed 10 April 2012).

72. Pereira WS, Freire RS. Ferro zero: Uma nova abordagem para o tratamento de águas contaminadas com compostos orgânicos poluentes. Química Nova 2005;28(1): 130-136. DOI: 10.1590/S0100-40422005000100022.

http://www.scielo.br/scielo.php?script=sci_arttext&pid=S0100-40422005000100022 (accessed 10 March 2012).

73. Houk VS. The genotoxicity of industrial wastes and effluents. Mutation Research 1992; 277 (2) 91-138.

74. Jager I, Hafner C, Schneider K. Mutagenicity of different textile dye products in Salmonella typhimurium and mouse lymphoma cells. Mutation Research 2004;561 (1-2) 35-44.

75. Chequer FMD, Angeli JPF, Ferraz ERA, Tsuboy MS, Marcarini JC, Mantovani MS, Oliveira DP. The azo dyes Disperse Red 1 and Disperse Orange 1 increase the micronuclei frequencies in human lymphocytes and in HepG2 cells. Mutation Research 2009; 676 (1-2) 83-86.

76. Chequer FMD, Lizier TM, Felicio R, Zanoni MVB, Debonsi HM, Lopes NP, Marcos R, Oliveira DP. Analyses of the genotoxic and mutagenic potential of the products formed after the biotransformation of the azo dye Disperse Red 1. Toxicology in Vitro 2011; 25 (8) 2054-2063

77. Ferraz ERA, Umbuzeiro GA, de-Almeida G, Caloto-Oliveira A, Chequer FMD, Zanoni MVB, Dorta DJ, Oliveira DP. Differential Toxicity of Disperse Red 1 and Disperse Red 13 in the Ames Test, HepG2 Cytotoxicity Assay, and Daphnia Acute Toxicity Test. Environmental Toxicology 2011; 26 (5) 489-497

78. Oliveira GAR, Ferraz ERA, Chequer FMD, Grando MD, Angeli JPF, Tsuboy MS, Marcarini JC, Mantovani MS, Osugi ME, Lizier TM, Zanoni, MVB, Oliveira DP. Chlorination treatment of aqueous samples reduces, but does not eliminate, the mutagenic effect of the azo dyes Disperse Red 1, Disperse Red 13 and Disperse Orange 1. Mutation Research 2010 703 (2) 200-208

79. Konstantinou IK, Albanis TA. TiO2-assisted photocatalytic degradation of azo dyes in aqueous solution: kinetic and mechanistic investigations - A review. Applied Catalysis B: Environmental 2004; 49 (1) 1-14. DOI:10.1016/j.apcatb.2003.11.010.

80. Hessel C, Allegre C, Maisseu M, Charbit F, Moulin P. Guidelines and legislation for dye house effluents. Journal of Environmental Management 2007; 83 (2) 171-180. DOI: 10.1016/j. jenvman. 2006.02.012.

81. Carneiro PA, Osugi ME, Fugivara CS, Boralle N, Furlan M, Zanoni MVB. Evaluation of different electrochemical methods on the oxidation and degradation of Reactive Blue 4 in aqueous solution. Chemosphere. 2005; 59(3) 431-439.

82. Mohan SV, Bhaskar YV, Karthikenyan J. Biological decolourization of simulated azo dye in aqueous phase by algae Spirogyra species. International Journal of Environment and Pollution. 2004; 21 (3) 211-222.

83. Vaghela SS, Jethva AD, Mehta BB, Dave SP, Adimurthy S, Ramachandraiah G. Laboratory studies of electrochemical treatment of industry azo dye effluent. Environmental Science and Technology. 2005; 39 (8) 2848-2855.

84. Cardoso JC, Lizier TM, Zanoni MVB. Highly ordered TiO2 nanotube arrays and photoelectrocatalytic oxidation of aromatic amine. Applied Catalysis, B: Environmental. Applied Catalysis, B: Environmental. 2010; 99 (1-2) 96-102.

85. Pereira RS. Identificação e caracterização das fontes de poluição em sistemas hídricos. Revista Eletrônica de Recursos Hídricos. IPH-UFRGS 2004; 1(1): 20-36. http://www.abrh.org.br/informacoes/rerh.pdf (accessed 06 June 2012).

86. Baumgarten, MGZ.; Pozza, SA. Qualidade de Águas. Descrição de Parâmetros Químicos referidos na Legislação Ambiental. Ed: FURG; 2001.

87. Prado AGS, Torres JD, Faria EA, Dias SCL. Comparative adsorption studies of indigo carmine dye on chitin and chitosan. Journal of Colloid and Interface Science. 2004; 1(1) 43-47.

88. Babel S, Kurniawan TA. Low-cost adsorbents for heavy metals uptake from contaminated water: a review. Journal of Hazardous Materials. 2003; 97(1-3) 219-243.

89. Associação Brasileira da Indústria Têxtil e de Confecção. Economia. <http://www.abit.org.br/site/navegacao.asp?id_menu=8&IDIOMA=PT> (accessed 19 August 2011).

90. Walsh FC. Electrochemical technology for environmental treatment and clean energy conversion. Pure and Applied Chemistry. 2001; 73 (12) 1819-1837.

91. Flores ER, Perez F, Torre M. Scale-up of Bacillus thuringiensis fermentation based on oxygen transfer. Journal of Fermentation and Bioengineering. 1997; 83 (6) 561-564.

92. Bornick H. Simulation of biological degradation of aromatic amines in river bed sediments. Water Research. 2001; 35 (3) 619-624.

93. Wang L, Barrington S, Kim JW. Biodegradation of pentylamine and aniline from petrochemical wastewater. Journal of Environmental Management. 2007; 83 (2) 191-197.

94. Orshansky F, Narkis N. Characteristics of organics removal by pact simultaneous adsorption and biodegradation. Water Research. 1997; 31(3) 391-398.

95. Fukushima M, Tatsumi K, Morimoto K. The fate of aniline after a photo-fenton reaction in an aqueous system containing iron(III), humic acid, and hydrogen peroxide. Environmental Science and Technology. 2010; 34 (10) 2006-2013.

96. Pramauro E. et al. Photocatalytic treatment of laboratory wastes containing aromatic amines. Analyst. 1995; 120 (2) 237-242.

97. Augugliaro V. et al. Photodegradation kinetics of aniline, 4-ethylaniline, and 4-chloroaniline in aqueous suspension of polycrystalline titanium dioxide. Research on Chemical Intermediates. 2000; 26 (5) 413-426.

98. Canle LM, Santaballa JA, Vulliet E. On the mechanism of TiO2-photocatalyzed degradation of aniline derivatives. Journal of Photochemistry and Photobiology, A: Chemistry. 2005; 175 (2-3) 192-200.

99. Chu W, Choy WK, So TY. The effect of solution pH and peroxide in the TiO2-induced photocatalysis of chlorinated aniline. Journal of Hazardous Materials. 2007; 141 (1) 86-91.

100. Low, G. K. C.; McEvoy, S. R.; Matthews, R. W. Formation of nitrate and ammonium ions in titanium dioxide mediated photocatalytic

degradation of organic compounds containing nitrogen atoms. Environmental Science and Technology, v. 25, n. 3, p. 460-467, 1991.

101. Pinheiro HM, Touraud E, Thomas O. Aromatic amines from azo dye reduction: status review with emphasis on direct UV spectrophotometric detection in textile industry wastewaters. Dyes and Pigments. 2004; 61(2) 121-139.

102. Fukushima M, Tatsumi K, Morimoto K. The Fate of Aniline after a Photo-Fenton Reaction in an Aqueous System Containing Iron(III), Humic Acid, and Hydrogen Peroxide. Environmental Science and Technology. 2006; 34 (10) 2006-2013.

103. Chu W, Choy WK, So TY. The effect of solution pH and peroxide in the TiO2-induced photocatalysis of chlorinated aniline. Journal of Hazardous Materials. 2007; 141 (1) 86-91.

104. Finklea HO. Semiconductor electrodes. New York: Elsevier; 1998.

105. Lin SH, Chen ML. Purification of textile wastewater effluents by a combined Fenton process and the ion exchange. Desalination 1997; 109 (1) 121-130.

106. Mozia S. Photocatalytic membrane reactor (PMRs) in water and wastewater treatment. A review. Separation and Purification Technology 2010; 73 (2) 71-91.

Chapter 4

The Use of New Technologies in Dyeing of Proteinous Fibers

Rıza Atav[1]

[1] Namık Kemal University, Textile Engineering Department, Turkey

1. INTRODUCTION

Wool consists principally of one member of a group of proteins called keratins; other members of this group include the proteins of hair, feathers, beaks, claws, hooves, horn and even certain types of skin tumour [1]. The main substance of wool is a keratin. Keratin macromolecules are crosslinked with cystine residues and contain a variety of side chains, some basic and some acidic [2]. Wool fibers consist of cortical cells and cuticular cells, which are located in the outermost part of the fiber surrounding the cortical cells. They consist of endocuticle, A and B exocuticle, and an exterior hydrophobic thin membrane called the epicuticle [3]. Both layers have a tremendous influence on dyeing because of their hydrophobic characteristics. The cuticle is separated from the underlying cortex by the intercellular material, which is called the cell membrane complex (CMC) and consists of non-keratinous proteins and lipids [2].

The morphology of the wool fiber surface plays an important role in textile finishing processes. The covalently bound fatty acids and the high amount of disulphide bridges make the outer wool surface highly hydrophobic.

Especially in the printing and dyeing of wool, the hydrophobic character of the wool surface is disturbing. Diffusion of the hydrophilic dyes at and into the fibers is hindered. For this reason, the hydrophilicity and dyeability properties of the wool fiber should be developed [4]. Wool dyeing is a degradative process involving high temperature for long periods in acidic to neutral pH medium to achieve good penetration, optimum fastness, and dye uptake. The results can be harsh handle, discomfort, and a deterioration of properties that impact consumer wear, care, and aesthetic appreciation [5].

Silk fiber is protein fiber that is produced from silk worms [6]. Silk has been called “the queen of fibers”, its natural luster, handle and draping properties being superior to those of many other textile fibers [7]. It is composed of different alpha amino acids orienting to form long chain polymer by condensation and polymerization. Silk fiber consists of 97% protein and the others are wax, carbohydrate, pigments, and inorganic compounds. The proteins in silk fiber are 75% fibroin and 25% sericin by weight, approximately. The sericin makes silk fiber to be strong and lackluster; therefore, it must be degummed before dyeing [6]. Silk fibroin, like wool keratin, is formed by the condensation of α-amino acids into polypeptide chains, but the long-chain molecules of silk fibroin are not linked together by disulfide bridges as they are in wool. Chemical treatments can cause modification of main peptide chains, and side chains of amino acids, which in turn influence the fiber’s chemical, physical, and mechanical properties [8]. Silk fiber is easily damaged when dyeing at the boil, so low-temperature dyeing is usualy preferred [7]. Because the brilliancy of dyed and printed silk fabrics is a decisive factor for evaluating the quality of silk fabrics, dyeability of silk fibers is one of the most attractive topics for applied and basic research [9].

In recent years, many attempts have been made to improve various aspects of dyeing, and new technologies have been, and are being developed to reduce fiber damage, decrease energy consumption and increase productivity [10]. In this chapter, new technologies that improve the dyeability of proteinous fibers such as ultrasound, ultraviolet, ozone, plasma, gamma irradiation, laser, microwave, e-beam irradiation, ion implantation, and supercritical carbondioxide will be overviewed.

2. ULTRASOUND TECHNOLOGY

Sound is transmitted through a medium by inducing vibrational motion of the molecules through which it is traveling. This vibrational motion represents the sound frequency [11]. Ultrasound is sound of a frequency that is above the threshold of human hearing [12]. The lowest audible frequency for humans is about 18Hz and the highest is normally around 18-20 kHz for adults, above which it becomes inaudible and is defined as ultrasound [11]. In recent decades the use of ultrasound technology has established an important place in different industrial processes such as the medical field, and has started to revolutionize environmental protection. The idea of using ultrasound in textile wet processes is not a new one. On the contrary there are many reports from the 1950s and 1960s describing the beneficial effects of ultrasound in textile wet processes. In spite of encouraging results from laboratory-scale studies, the ultrasound-assisted wet textile processes have not been implemented on an industrial scale as yet [13].

In practice, three ranges of frequencies (Fig. 1) are reported for three distinct uses of ultrasound: low frequency or conventional power ultrasound (20-100 kHz), medium frequency ultrasound and diagnostic or high frequency ultrasound (2-10 MHz) [13].

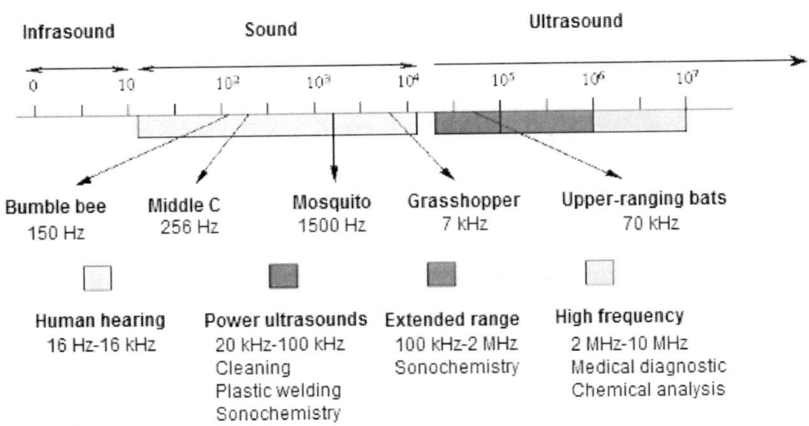

Figure 1. Classification of sound according to the frequency [**14**]

As the sound wave passes through water in the form of compression and rarefaction cycles, the average distance between the water molecules varies. If the pressure amplitude of the sound is sufficiently large, then the distance between the adjacent molecules can exceed the critical molecular distance during the rarefaction cycle. At that moment a new liquid surface is created in the form of voids. This phenomenon is called acoustic cavitation [15].

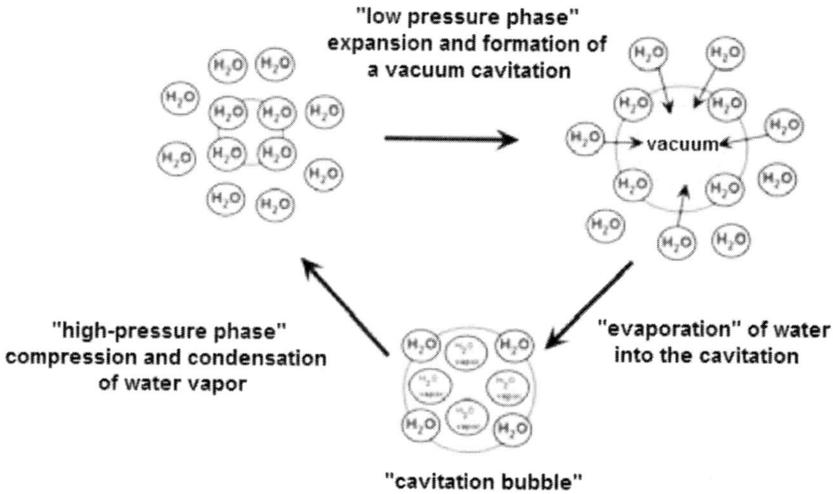

Figure 2. Formation of a cavitation buble [**16**]

Power ultrasound can enhance a wide variety of chemical and physical processes, mainly due to the phenomenon known as cavitation in a liquid medium that is the growth and explosive collapse of microscopic bubbles. Sudden and explosive collapse of these bubbles can generate ''hot spots'', i.e., localized high temperature, high pressure, shock waves and severe shear force capable of breaking chemical bonds [17]. High temperature and pressures resulting from the collapse of the transient cavitation bubbles are responsible for all the observed effects of ultrasound [18]. Parameters which affect cavitation and bubble collapse are:

- Properties of the solvent: The solvent used to perform sample treatment with ultrasonication must be carefully chosen. As a general rule, most applications are performed in water. However, other less polar liquids, such as some organics, can be also used, depending on the intended purpose [19]. Cavities are more readily formed when using a solvent with high vapor pressure, low viscosity and low surface tension. But at high vapor pressure more vapor enters the cavitation bubble during its formation and the bubble collapse is cushioned and less violent [13].

- Properties of gases: Soluble gases should result in the formation of a larger number of cavitation nuclei, but the greater the solubility of the gas is the more gas molecules should penetrate the cavity. Therefore, a less violent and intense shock wave is created on bubble collapse [13].

- External pressure: With increasing external pressure, the vapor pressure of the liquid decreases and higher intensity is necessary to induce cavitation [13]. In addition, there is an increment in the intensity of the cavitational bubble collapse and, consequently, an enhancement in sonochemical effects is obtained. For a specific frequency there is a particular external pressure that will provide an optimum sonochemical reaction [19].

- External temperature: Higher external temperature reduces the intensity necessary to induce cavitation due to the increased vapor pressure of the liquid. At higher external temperatures more vapor diffuses into the cavity, and the cavity collapse is cushioned and less violent [13].

- Frequency of the sound wave: At high sonic frequencies, on the order of the MHz, the production of cavitation bubbles becomes more difficult than at low sonic frequencies, of the order of the kHz. To achieve cavitation, as the sonic frequency increases, so the intensity of the applied sound must be increased, to ensure that the cohesive forces of the liquid media are overcome and voids are created [19]. Lower frequency produces more violent cavitation and, as a consequence, higher localized temperatures and pressures. At very high frequency, the expansion part of the sound wave is too short to permit molecules to be pulled apart sufficiently to generate a bubble [13].

Cavitation induced by ultrasound will allow accelerating processes and obtaining the same results as existing techniques but with a lower temperature and low dye and chemical concentrations [20]. For this reason textile wet processes assisted by ultrasound are of high interest for the textile industry. A review of earlier studies using ultrasound in textile wet processes was compiled by Thakore et al. Ultrasound-assisted textile dyeing was first reported by Sokolov and Tumansky in 1941 [13]. Some of the benefits of using of ultrasonics in dyeing can be listed as below;

- energy savings by dyeing at lower temperatures and reduced processing times,
- environmental improvements by reduced consumption of auxiliary chemicals,
- lower overall processing costs (due to less energy and chemical consumption), thereby increasing industry competitiveness [21]
- Improvements observed in ultrasound-assisted dyeing processes are generally attributed to 3 main phenomenons,
- Dispersion: breaking up of micelles and high molecular weight aggregates into uniform dispersions in the dye bath,
- Degassing: expulsion of dissolved or entrapped gas or air molecules from fiber into liquid and removal by cavitation, thus facilitating dye-fiber contact, and
- Diffusion: accelerating the rate of dye diffusion inside the fiber by piercing the insulating layer covering the fiber and accelerating the interaction or chemical reaction, if any, between dye and fiber [11].

A good wool dyeing process must provide a satisfactory uptake of dye bath and an adequate penetration of dye into the fiber, with the practical advantages of good wet fastness and uniform coloration. The conventional methods for wool dyeing are based on long times at temperature close to the boiling point, in order to ensure good results of dye penetration and leveling. These conditions can damage the fibers, with negative effects on the characteristics of the finished material. Such damage can be minimized by reducing the operation time or, better yet, by reducing the dyeing temperature. Recently, ultrasound assisted wool dyeing was studied with the aim to reduce temperature or dyeing time with respect to the conventional dyeing technique [22]. Some literature related to the use of ultrasound technology in dyeing of proteinous fibers is summarized below.

Shukla and Mathur (1995) studied the dyeing process of silk using cationic, acid and metal-complex dyes at low temperatures, assisted by a low frequency ultrasound of 26 kHz and compared the results of dye uptake with those obtained by conventional processes. Their results show that silk dyeing in the presence of ultrasound increases the dye uptake for all classes of dyes at lower dyeing temperatures (45°C and 50°C) and a shorter dyeing time (15 min.), as compared with conventional dyeing at 85°C for 60 min. Furthermore, there was no apparent fiber damage caused by cavitation [23].

Kamel et al. (2005) have investigated the dyeing of wool fabrics with lac as a natural dye in both conventional and ultrasonic techniques. The extractability of lac dye from natural origin using power ultrasonic was also evaluated in comparison with conventional heating. The results of dye extraction indicate that power ultrasonic is rather effective than conventional heating at low temperature and short time. Color strength values obtained were found to be higher with ultrasonic than with conventional heating. The results of fastness properties of the dyed fabrics were fair to good [17].

Vankar and Shanker (2008) have extracted coloring pigment from Hollyhock (Alcea rosea) flower and used for dyeing wool yarn, silk and cotton fabrics. It is observed that the dyeing with hollyhock gives fair to good fastness properties in sonicator in 1 hour and shows good dye uptake as compared with conventional dyeing [24].

Battu et al. (2010) observed that in wool dyeing at 85°C with acid dyes, ultrasound caused an improvement of the dye uptake as much as 25%, or dyeing time would be nearly 20% shorter than conventional dyeing [25].

Yukseloğlu and Bolat (2010) stated that the wool fabrics have presented similar color yield (K/S) and acceptable color differences (ΔE) with the use of ultrasonic energy. Ultrasonic energy was found to be advantageous to be used for wool dyeing at lower temperatures (such as 80°C and 90°C) and lower dyeing times (i.e. 80 min. or 90 min.) as an alternative process for conventional dyeing (100°C and 144 min) [26].

McNeil and McCall (2011) investigated the effects of ultrasound at 35-39 kHz on several wool dyeing and finishing processes. Ultrasound pre-treatment increased the effectiveness of subsequent oxidative-reductive bleaching, but had no effect on the uptake of acid leveling and acid milling dyes. The pre-treatment retarded the uptake of reactive dye, possibly by increasing the crystallinity of the fiber or removing surface bound lipids. Ultrasound did not improve dyeing under conditions that are currently used in industry, but did show potential to reduce the chemical and energy requirements of dyeing wool with reactive and acid milling dyes, but not acid leveling dyes [27].

Atav and Yurdakul (2011) investigated the effect of ultrasound usage on the color yield in dyeing of mohair fibers. They found that dyeing in the presence of ultrasound energy increases the dye-uptake of mohair fibers and hence higher color yield values are obtained. The difference between the samples dyed in the presence and absence of ultrasound, was greater for darker shades and for dyeing carried out in acidic medium (pH 5), and also for shorter dyeing periods. Furthermore there is no important difference between washing fastness and alkali solubility values of fibers dyed in the presence and absence of ultrasound [28].

Ferrero and Periolatto (2012) studied the possibility of reducing the temperature of conventional wool dyeing with an acid leveling dye using ultrasound in order to reach dye uptake values comparable to those obtained with the standard procedure at 98°C. Dyeings of wool fabrics were carried out in the temperature range between 60°C and 80°C using either mechanical or ultrasound agitation of the bath and coupling the two methods to compare the results. For each dyeing, the dye uptake curves of the dye bath were determined and the better results of dyeing kinetics were obtained with ultrasound coupled with mechanical stirring. Finally, fastness tests to rubbing and domestic laundering yielded good values for samples dyed in ultrasound assisted process even at the lower temperature [22].

3. ULTRAVIOLET TECHNOLOGY

Light is electromagnetic radiation or radiant energy traveling in the form of waves [29]. The electromagnetic spectrum is the range of all possible frequencies of electromagnetic radiation. The "electromagnetic spectrum" of an object is the characteristic distribution of electromagnetic radiation emitted or absorbed by that particular object. The electromagnetic spectrum extends from low frequencies used for modern radio communication to gamma radiation at the short-wavelength (high-frequency) end, thereby covering wavelengths from thousands of kilometers down to a fraction of the size of an atom [30]. UV energy is found in the electromagnetic spectrum between visible light and x-rays [29].

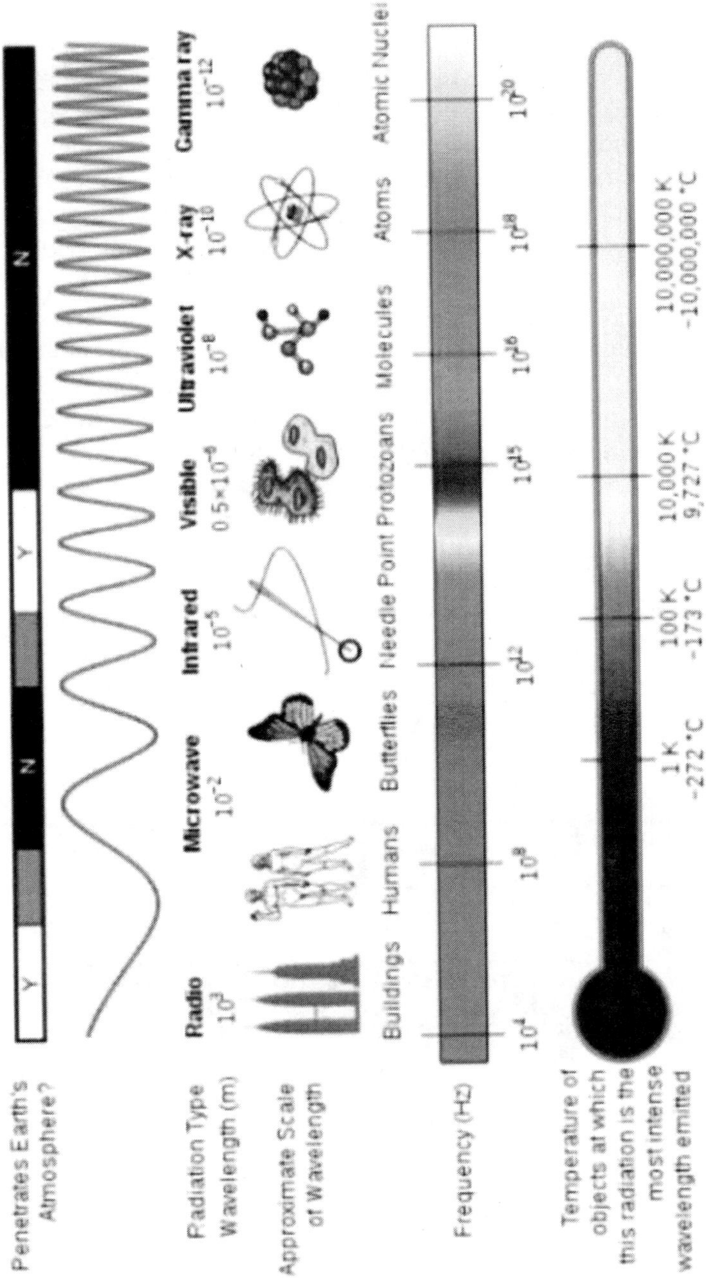

Figure 3. A diagram of the electromagnetic spectrum [**30**]

Ultraviolet or UV radiation is part of the electromagnetic (light) spectrum that reaches the earth from the sun. It has wavelengths shorter than visible light, making it invisible to the naked eye [31]. Ultraviolet radiation constitutes to 5% of the total incident sunlight on earth surface (visible light 50% and IR radiation 45%). Even though, its proportion is quite less, it has the highest quantum energy compared to other radiations [32]. Scientists classify UV radiation into three types or bands: UVA, UVB, and UVC (Fig. 4). The ozone layer absorbs some, but not all, of these types of UV radiation [33].

- UVA: Long-wavelength UVA covers the range 315-400 nm. Not significantly filtered by the atmosphere. Approximately 90% of UV radiation reaching the Earth's surface. UVA is again divided into UVA-I (340 nm - 400 nm) and UVA-II (315 nm - 340 nm) [34].

- UVB: Medium-wavelength UVB covers the range 280-315 nm. Approximately 10% of UV radiation reaching the Earth's surface [34].

- UVC: Short-wavelength UVC covers the range 100-280 nm [34]. They are the most dangerous among all the rays. However, these rays do not reach the earth's surface as they are completely absorbed by the ozone layer [31].

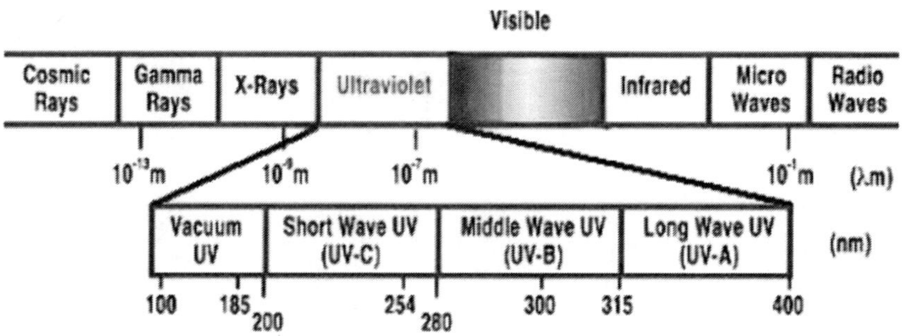

Figure 4. Classification of ultraviolet radiation [**35**]

The deleterious effects of solar irradiation are perceived as changes in texture and color, dryness, etc., and can be evaluated in terms of reduced elasticity, increased porosity or swelling properties, altered dye sorption characteristics, and photofading of natural or artificial hair color [36]. The chemical changes caused by short-term UV irradiation of wool are confined to fibers at the fabric surface and UV is unable to penetrate beyond the surface to weaken the bulk fibers responsible for the mechanical strength. This has enabled the potential application of UV technology as a surface-specific treatment in several areas of wool processing [37]. For wool the UV-absorbing species are aromatic amino acid and cystine residues in the protein structure which absorb strongly below 350 nm. UVC radiation (200-280 nm) is the most effective range for modifying wool fiber surfaces [38].

The most commercially used process for antifelting and antipilling of wool is based on chlorination. However, recent concern over the release into the environment of adsorbable organohalogens (AOX) in process effluents has prompted the development of alternative, AOX-free processes. Different types of radiation techniques, such as ultraviolet radiation, are utilized as alternatives to chlorination in wool processing [39]. UV treatment can add value in coloration (dyeing and printing), since it is predominantly surface fibers in a fabric that absorb, reflect and scatter light. Photomodification of the surface fibers can allow:

- more dye to become fixed, producing deeper shades
- more rapid fixation of dyes dye fixation under less severe conditions (e.g. lower temperature) [38]

Modification of the dye uptake by exposure of wool fabric to UV radiation before dyeing has been known since the early 1960s. For most dye classes, UV-irradiated fabric takes up significantly more dye than untreated and when fabrics are irradiated through stencils, intricate tone-ontone effects can be produced [37]. Some literature related to the use of ultraviolet technology in dyeing of proteinous fibers is summarized below.

Millington (1998) stated that UV irradiation of wool can significantly increase dyeing color yields. The use of 1:1 metal-complex dyes was found to be particularly effective, and a 3% o.w.f. dyeing on UV-treated fabric could produce a better depth of shade than a 5% dyeing on untreated fabric [40].

Millington (1998) found that UV radiation of wool fabric exhibits some physical and chemical changes on its surface. This interaction not only modifies the fabric of wool but also improve the shades particularly grey and black. It also helps in even dyeing, deeper shades, chlorine free printing and improve the photo bleaching of wool [41].

Xin et.al (2002) exposed wool samples to UV radiation for 60 min. and investigated the surface modification of the wool fiber by X-ray photoelectron spectroscopy. The chemical change caused by the UV treatment was identified as surface oxidation of cystine (disulphide bonds) and thereby induced changes in the dyeing properties of the wool. The dyeability of UV-treated and untreated wool samples was determined at temperatures of 45, 50, 55 and 60°C using C.I. Acid Blue 7. The UV-treated wool samples showed greater levels of dye uptake compared with those of the untreated samples. The adsorption behavior and diffusion coefficients were also studied. The dyeing properties of wool were enhanced by UV radiation due to the increased diffusion coefficient of the dyes in the treated wool fibers [42].

4. OZONE TECHNOLOGY

Ozone is a natural occurring gas that can be both beneficial and detrimental to organisms on Earth. It is important that sufficient amount of this pale blue gas is present in the stratosphere, where O_3molecules would shield most of the UV radiation from reaching Earth [43]. Although ozone is a blue colored gas at normal temperatures and pressures; because of its low concentrations in its applications the observation of this blue color of ozone is impossible [44]. Ozone gas has a pungent odor readily detectable at concentrations as low as 0.02 to 0.05 ppm (by volume), which is below concentrations of health concern [45].

Ozone was first generated and characterized by a German scientist named Schonbein in 1840 [46]. Ozone is a nonlinear triatomic molecule possessing two interoxygen bonds of equal length (1.278 A) and an average bond angle of 116°49′ [47].

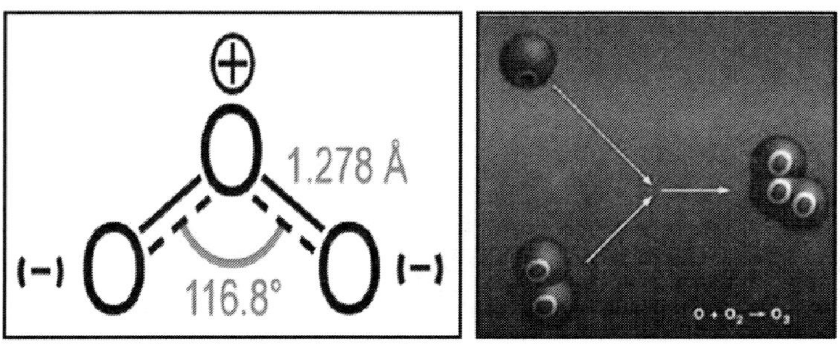

Figure 5. Ozone molecule [**48**, **49**]

Ozone is formed naturally in the atmosphere by photochemical reaction with solar UV radiation and by lightening. It can also be generated artificially. Three most common ways of generating ozone artificially are:

- Corona discharge: In this method, ozone is generated when free, energetic electrons in the corona dissociate oxygen molecules in oxygen-containing feed gas that passes through the discharge gap of the ozone generator.

- UV light: Ozone can also be generated by UV light. The high energy UV light ruptures the oxygen molecules into oxygen atoms, and the subsequent combination of an oxygen atom with an oxygen molecule produces ozone (O_3).

- Electrolysis: A third method for generating ozone is electrolysis, which uses an electrolytic cell. Specifically, electrolysis involves converting oxygen in the water to ozone by passing the water through positively and negatively charged surfaces [46].

Ozone is a very powerful oxidizing agent, which is able to participate in a great number of reactions with organic and inorganic compounds. Among the most common oxidizing agents, it is only surpassed in oxidant power by fluorine and hydroxyl radicals (see Table 1) [50]. Ozone has strong tendency to react with almost any organic substance as well as with water. The reaction proceeds via several intermediates such as peroxides, epoxides and perhydroxyl and hydroxyl radicals [51].

In an ozonation process two possible pathways have to be considered [50]: direct oxidation with ozone molecules or the generation of free-radical intermediates, such as the OH radical, which is a powerful, effective, and non-selective oxidizing agent [52].

$$O3+ H2O \rightarrow 2HO2 \qquad (1)$$

$$O3+ HO2\cdot \rightarrow HO\cdot + 2O2 \qquad (2)$$

Molecular ozone can directly react with dissolved pollutants by electrophilic attack of the major electronic density positions of the molecule. This mechanism will take place with pollutants such as phenols, phenolates or tiocompounds. The radical mechanism predominates in less reactive molecules, such as aliphatic hydrocarbons, carboxylic acids, benzenes or chlorobenzenes [50].

Table 1. Oxidation power of selected oxidizing species [**50**]

Oxidation species	Oxidation power (V)	Oxidation species	Oxidation power (V)
Fluorine	3.03	Chlorine dioxide	1.50
Hydroxyl radical	2.80	Hypochlorous acid	1.49
Atomic oxygen	2.42	Hypoiodous acid .	1.45
Ozone	2.07	Chlorine	1.36
Hydrogen peroxide	1.77	Bromide	1.09
Permanganate	1.67	Iodine	0.54
Hypobromous acid	1.59		

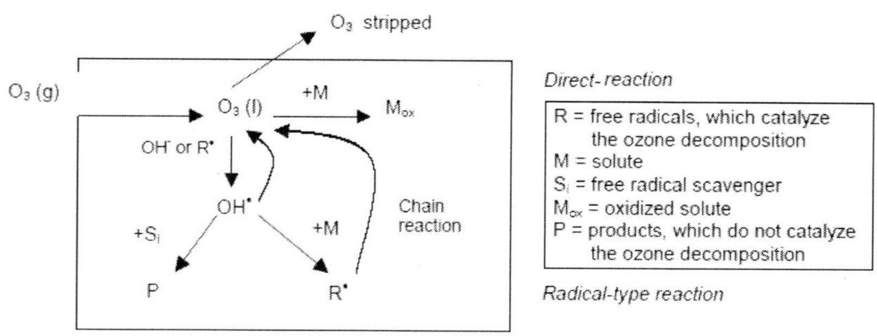

Figure 6. Scheme of reactions of ozone added to an aqueous solution [**50**]

The chlorine/ Hercosett process, the most widely used treatment for the wool dyeing process, causes dangerous ecological problems due to the contamination of waste water with absorbable organic halides (AOX). Because of legal restrictions and national and international awareness of ecology and pollution control, an AOX-free pretreatment is required to offer environmental advantages [53]. Alternative surface modifications for improving wool dyeability are therefore being explored. One of them is ozonation process [54]. Ozone treatments of proteinous fibers such as wool, mohair, angora and silk have been investigated by many authors. When literature is examined, it can be well understood that, increase in dyeability of protenious fibers caused by ozonation process depends on the following parameters;

- **pH:** Typically, at pH<4 direct ozonation dominates and above pH>9 the indirect pathway dominates. In the range of pH 4-9, both of them are important. The pH influences the generation of hydroxyl radicals [55].

$$O3+ OH- \rightarrow O3-\bullet + OH\bullet \quad (3)$$

$$O_3^{-}\bullet \rightarrow O^{-}\bullet + O_2 \quad (4)$$

$$O^{-}\bullet + H^{+} \rightarrow OH\bullet \quad (5)$$

Generally at neutral medium reaction rate of ozone gas is slow due to its low solubility. While molecular ozone reacts at low pH values, at high pH values radicals react. Since the oxidation potential of hydroxyl radicals exceeds that of ozone molecules, oxidation is faster in indirect reactions. Additionally HO• is not the only radical that is formed. Even though the HO• radical is the most powerful radical with a 2.8 V oxidation potential, $HO_2\bullet$, $HO_3\bullet$ and $HO_4\bullet$ radicals are also formed as shown in Eq. 6-11 [56].

$$O_3 + OH^{-} \rightarrow HO_2\bullet + O_2^{-} \quad (6)$$

$$HO_2\bullet \leftrightarrow O_2^{-}\bullet + H^{+} \quad (7)$$

$$O_3 + O_2^{-}\bullet \rightarrow O_3^{-} + O_2 \quad (8)$$

$$O_3^{-} + H^{+} \rightarrow HO_3 \quad (9)$$

$$HO_3\bullet \rightarrow HO\bullet + O_2 \tag{10}$$

$$HO\bullet + O_3 \rightarrow HO_4 \tag{11}$$

At pH < 9, the highly selective ozone molecules react rapidly at sites of high electron density (such as aliphatic and aromatic double bonds) and slowly at less reactive sites (such as C-H bonds of saturated hydrocarbons). The presence of the •OH radicals above pH 9 have less selectivity and high oxidation potential (2.8 V) [51]. Ozonation efficiency decreases in the basic pH values when compared to acidic and neutral pH. Because in a basic solution, more hydroxide ions are present and these hydroxide ions act as an initiator for the decay of ozone [57]. Moreover, it is reported that the dissolved ozone concentration in water decreases from $4.3\text{X}10^{-4}$ mol/L at pH 4 to $1.5\text{X}10^{-4}$ mol/L at pH 10 [58].

- **Temperature:** As temperature increases, ozone becomes less soluble (see Table 2) [45]; however, it can not be said that ozonation efficiency reduces with the decrease in the solubility of ozone, because temperature rise also increases the reaction rate [56].

Table 2. Water solubility of ozone [**59**]

Temperature (°C)	Solubility ($kg.m^{-3}$)
0	1.09
10	0.78
20	0.57
30	0.40
40	0.27
50	0.19
60	0.14

- **Ozone dose:** Because oxidation reactions are caused by molecular ozone or radical species formed by the reactions of ozone, with the increase in ozone dose ozonation efficiency increases [56]

- **Water content of the fiber:** In literature it was stated that the rate of oxidation is accelerated by the hydration of hydrophilic groups in the fiber and above a critical level of moisture content (at which fiber is completely hydrated) it is retarded by the water which enters the intermicellar and interfibrillar space. In many previous studies, the importance of water during ozonation process was reported [51, 57, 58]. Prabaharan and Rao have suggested a model in order to explain this phenomenon. According to this model it can be said that for ozonation to take place, water should be present inside of the fiber. However the quantity of water present has a definite effect on the rate of reaction [51].

Fig. 7 gives the schematic representation of ozone path from gas phase to reaction site in the fiber. When sufficient water is not present, d_1 and d_2 are absent and hence O is transported by convection across the distances d_1 and d_2 and then by diffusion across d_3 and d_4. Since sufficient water is not supplied, entire hydrophilic group in the fiber is not hydrated and hence the extent of attack at R is low. When sufficient water is supplied, d_1 is absent and d_2 is either fully or partially absent depending on the quantity of water available and hence O is transported by convection across d_1 and convection or diffusion across d_2 followed by diffusion across d_3 and d_4. Since sufficient water is supplied, entire hydrophilic group in the fiber is hydrated and hence the extent of attack at R is maximum. When excess water is present, d_1 and d_2 are present and hence O is transported by diffusion across d_1, d_2, d_3, and d_4. Since excess water is present at d_1 and d_2, dilution of ozone takes place and hence ozone attack at R is lower in spite of complete hydration at d_3 [58].

Some literature related to the effect of ozonation on dyeing properties of various proteinous fibers is summarized below.

Micheal and El-Zaher (2003) has evaluated the effect of ultraviolet/ozone treatments for different times on the characteristics of wool fabrics with respect to wettability, permeability, yellowness index, and weight loss. The beneficial effects of this treatment on dyeability, color parameters, light fastness characteristics, and the change in color difference after exposure of the treated dyed samples to artificial daylight for about 150 hours were investigated. The results indicated that the improvement in wetting processes may have been due to surface modifications; this meant that an increase in the amorphousity of the treated samples, the oxidation of the cystine linkage on the surface of the fabrics, and the formation of free-radical species encouraged dye uptake [60].

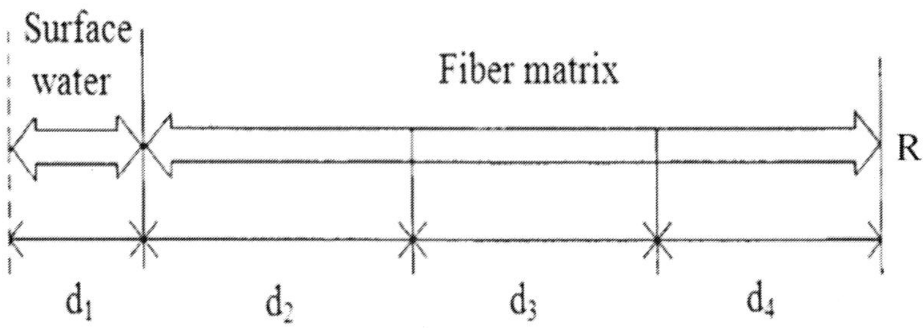

Figure 7. Schematic representation of ozone path from gas phase to reaction site [**58**]; (d_1: the distance occupied by surface water, d_2: the distance occupied by mobile water phase, d_3: immobile water phase, d_4: the distance between the immobile water phase and R)

Sargunamani and Selvakumar (2007) investigated the effects of process parameters (pick-up value, pH and time) in the ozone treatment of raw and degummed tassar silk fabrics on their properties such as yellowness index, breaking strength, breaking elongation, weight, amino group content. Decrease in yellowness index, breaking strength, breaking elongation, and weight as well an increase in amino group content was observed [58].

Perincek et.al (2008) investigated a novel bleaching technique for Angora rabbit fiber. For this purpose, a detailed investigation on the role of the fiber moisture, pH, and treatment time during ozonation was carried out. Also, the effect of ozonation on the dyeing properties of Angora rabbit fibers was researched. Consequently, it was found that ozonation improved the degree of whiteness and dyeability of Angora rabbit fiber [57].

Atav and Yurdakul (2010) investigated the use of ozonation to achieve dyeability of the angora fibers at lower temperatures without causing any decrease in dye uptake by modifying the fiber surfaces. The study was carried out with known concentration of ozone, involving process parameters such as wet pick-up (WP), pH, and treatment time. The effect of fiber ozonation was assessed in terms of color, and test samples were also evaluated using scanning electron microscopy (SEM). The optimum conditions of ozonation process were determined as WP 60%, pH 7 and 40 min. According to the experimental results it can be concluded that, ozonated angora fibers can be dyed at 90°C with acid and reactive dye classes without causing any decrease in color yield [61]. In an other study on ozonation process carried out by *Atav and Yurdakul (2011)* the optimum conditions for mohair fiber were determined as WP 60%, pH 7 and 30 min. Dyeing kinetics also studied and it was demonstrated that the rate constant and the standard affinity of ozonated sample increased [54].

5. PLASMA TECHNOLOGY

Faraday proposed to classify the matter in four states: solid, liquid, gas and radiant. Researches on the last form of matter started with the studies of Heinrich Geissler (1814-1879): the new discovered phenomena, different from anything previously observed, persuaded the scientists that they were facing with matter in a different state. Crookes took again the term "radiant matter" coined by Faraday to connect the radiant matter with residual molecules of gas in a low-pressure tube. Sufficient additional energy, supplied to gases by an electric field, creates plasma [62]. The plasma is referred to as the fourth state of matter (in addition to solid, liquid, gaseous) [63].

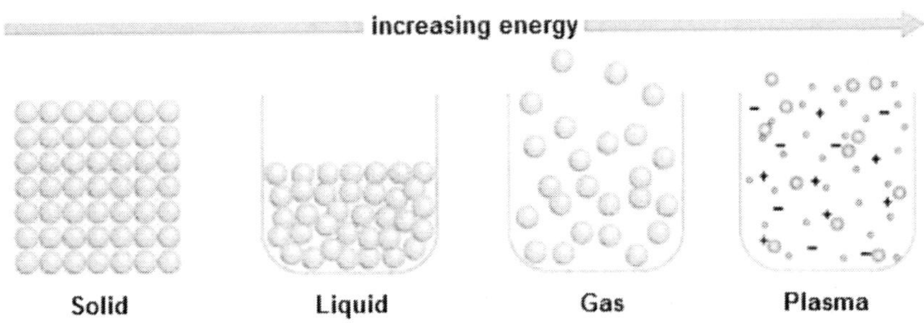

Figure 8. States of matter [**64**]

The plasma is an ionized gas with equal density of positive and negative charges which exist over an extremely wide range of temperature and pressure [65]. As shown in Fig. 9, the plasma atmosphere consists of free electrons, radicals, ions, UV-radiation and a lot of different excited particles in dependence of the used gas [66].

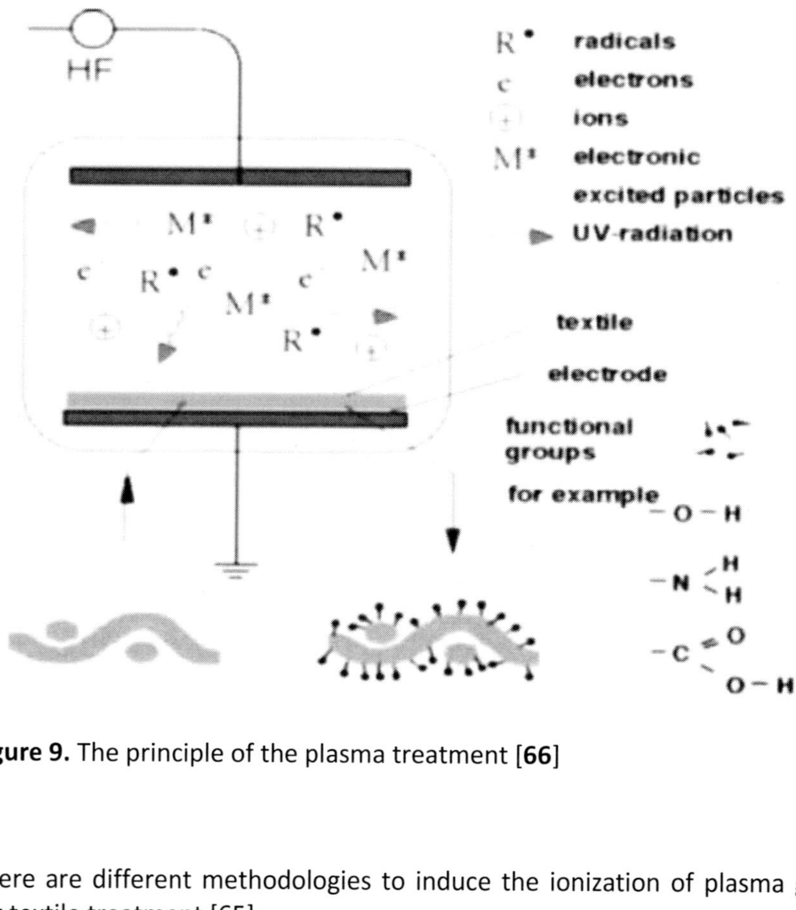

Figure 9. The principle of the plasma treatment [**66**]

There are different methodologies to induce the ionization of plasma gas for textile treatment [65]:

- Glow Discharge: It is the oldest type of plasma; it is produced at reduced pressure and assures the highest possible uniformity and flexibility of any plasma treatment [67]. The methodology applies direct electric current, low frequency over a pair of electrodes [65]. Alternatively, a vacuum glow discharge can be made by using microwave (GHz) power supply [68].

- Corona Discharge: It is formed at atmospheric pressure by applying a low frequency or pulsed high voltage over an electrode pair [65]. Typically, both electrodes have a large difference in size. The corona consists of a series of small lightning-type discharges. High local energy levels and problems related to the homogeneity of the classical corona treatment of textiles make it problematic in many cases [68].

- Dielectric-Barrier Discharge: DBD is produced by applying a pulsed voltage over an electrode pair of which at least one is covered by a dielectric material [65]. Although lightning-type discharges are created, a major advantage over corona discharges is the improved textile treatment uniformity [68].

Practically, one generates the plasma by applying an electrical field over two electrodes with a gas in between. This can be carried out at atmospheric pressure or in a closed vessel under reduced pressure. In both cases, the properties of the plasma will be determined by the gasses used to generate the plasma, as well as by the applied electrical power and the electrodes (material, geometry, size, etc.). The pressure of the gas will have a large influence on the plasma properties but also on the type of equipment needed to generate the plasma [69].

The plasmas can be classified as being of the low pressure and atmospheric type. Both plasmas can be used for the surface cleaning, surface activation, surface etching, cross linking, chain scission, oxidation, grafting, and depositing of materials, and generally similar effects are obtained; however, atmospheric plasma has many advantages when compared with vacuum plasma [70]. *Low pressure plasmas* are typically in the pressure range of 0.01 kPa. A vacuum chamber and the necessary vacuum pumps are required, which means that the investment cost for such a piece of equipment can be high. These plasmas are characterized by their good uniformity over a large volume. *Atmospheric plasmas*operate at standard atmospheric pressure (~ 100 kPa). Open systems using the surrounding air exist. The range of processes is not as wide as for low pressure plasmas. On the other hand, these systems are easily integrated in existing finishing lines, a major advantage from industrial view point. Of course, for an inline process to be feasible, the plasma treatment has to be done at sufficiently high line speeds, which is not evident for textile materials [69].

Due to increasing requirements on the finishing of textile fabrics, increasing use of technical textiles with synthetic fibers, as well as the market and society demand for textiles that have been processed by environmentally sound methods, new innovative production techniques are demanded [66]. Plasma technology is an important alternative to wet treatments, because there is no water usage, treatment is carried out in gas phase, short treatment time is enough, it does not cause industrial waste, and it provides energy saving [71]. In Cold plasma, that is, low-temperature plasma (LTP) treatment is the most commonly used physical method for a surface specific fiber modification, as it affects the surface both physically and chemically [72]. The advantage of such plasma treatments is that the modification turns out to be restricted in the uppermost layers of the substrate, thus not affecting the overall desirable bulk properties [62].

The plasma gas particles etch on the fabric surface in nano scale so as to modify the functional properties of the fabric. Unlike conventional wet processes, which penetrate deeply into fibers, plasma only reacts with the fabric surface and does not affect the internal structure of the fibers. Plasma technology modifies the chemical structure as well as the topography of the textile material surface. In conclusion, plasma can modify the surface properties of textile materials, deposit chemical materials (plasma polymerization) to add functionality, or remove substances (plasma etching) from the textile materials [65]. Essentially, four main effects can be obtained depending on the treatment conditions;

- The cleaning effect: It means the removal of impurities or substrate material from the exposed surface [73]. It is mostly combined with changes in the wettability and the surface texture. This leads for example to an increase in dye-uptake.

- Increase of microroughness: This affects, for example, an anti-pilling finishing of wool.

- Generation of radicals: Presence of free radicals induces secondary reactions like cross-linking. Furthermore, graft polymerization can be carried out as well as reaction with oxygen to generate hydrophilic surfaces [66].

- Plasma polymerization: In plasma polymerization, a monomer is introduced directly into the plasma and the polymerization occurs in the plasma itself [73]. It enables the deposition of solid polymeric materials with desired properties onto the substrates [66].

When a surface is exposed to plasma a mutual interaction between the gas and the substrate takes place. The surface of the substrate is bombarded with ions, electrons, radicals, neutrals and UV radiation from the plasma while volatile components from the surface contaminate the plasma and become a part of it. Whatever may be the final outcome on the surface, the basic effect that causes modification is based on the *radical formation* (attachment of functional group and deposition/polymerization) and *etching* phenomena. Fig. 10 illustrates the mechanism of plasma modification [74].

Low temperature plasma treatment of wool has emerged as one of the environmental friendly surface modification method for wool substrate. The efficiency of the low temperature plasma treatment is governed by several operational parameters like;

- Nature of the gas used
- System pressure
- Discharge power
- Duration of treatment

Plasma treatment can impart anti-felting effect, degreasing, improved dyestuff absorption and increased wetting properties to wool fibers [68]. These effects of the plasma process are attributed to several changes in the wool surface, such as;

- the formation of new hydrophilic groups (sulphonate and carboxylate),
- partial removal of covalently bonded fatty acids belonging to the outermost surface of the fiber, and
- the etching effect [75].

Plasma treatments modify the fatty acid monolayer present in the outermost part of the wool fiber, generating new hydrophilic groups as a result of the hydrocarbon chain oxidation and reducing the fatty acid chain length. The oxidation process also promotes the formation of Bunte salt and cysteic acid residues on the polypeptide chain. Particularly when oxidizing gasses are used, plasma induces cystine oxidation in the A-layer of the exocuticle, converting it into cysteic acid and thus reducing the number of crosslinkages in the fiber surface [72]. As the surface is oxidized, the hydrophobic character is changed to become increasingly hydrophilic [76]. The etching of the hydrophobic epicuticle and increase in surface area also contributes towards the improvement in the ability of the fibers to wet more easily [74].

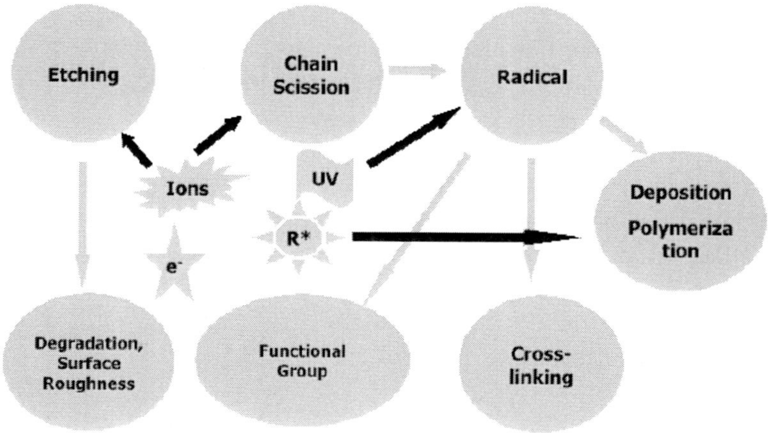

Figure 10. Mechanism of plasma-substrate interaction [**74**]

Plasma treatment of wool fibers causes improvement in dyeability of wool fibers due to the changes occurred on fiber surface. For this reason dyeing kinetics, dye uptake and hence depth of shade are increased. In literature there are many studies related to the effect of plasma treatment on dyeability of proteinous fibers. Some of them are summarized below.

Wakida et al. (1993) treated the merino wool top with low-temperature plasmas of helium/argon and acetone/argon under atmospheric pressure for 30 seconds and then dyed with two leveling acid dyes, and two milling acid dyes. Dyeing rate and final dye uptake increased with the atmospheric low-temperature plasma treatments. In particular, helium/argon plasma was found to be much more effective than acetone/argon plasma at improving dyeing properties [77].

Yoon et al. (1996) treated the wool fabrics with low temperature oxygen plasma and examined their mechanical and dyeing properties. Plasma pretreatment caused an increase in strength. Furthermore, it was observed that when wool was dyed with a leveling acid dye, equilibrium dye uptake did not change, but the dyeing rate increased with a milling acid dye [78].

Jing (1996) investigated the surface modification of silk fabric by plasma graft copolymerization with acrylamide and acrylic acid. The dependence of graft degree was examined on the conditions of plasma grafting. The relationships were discussed between graft degree and factors such as crease recovery, dyeability, colour fastness and mechanical properties. It was shown that the dyeability and color fastness have been improved for samples grafted with acrylic acid [79].

Wakida et al. (1996) treated wool fibers with oxygen low-temperature plasma and then dyed with acid and basic dyes. Despite the increase of electronegativity of the fiber surface caused by the plasma treatment, the rate of dyeing of wool was increased with both dyes [80].

Wakida et al. (1998) treated wool fabrics with oxygen, carbon tetrafluoride, and ammonia low temperature plasmas and then dyed with several natural dyes. The dyeing rate of the plasma-treated wool increased considerably with cochineal, Chinese cork tree, and madder, but not with gromwell. Furthermore, plasma-treated wool fabrics dyed with cochineal and Chinese cork tree have increased brightness compared with untreated wool [81].

Kan et al. (1998) investigated the induced surface properties of wool fabrics created by the sputtering of low-temperature plasma treatment, such as surface luster, wettability, surface electrostatic and dyeability. After the low-temperature plasma treatment, the treated wool fabric specimens exhibited better hydrophilicity and surface electrostatic properties at room temperature, together with improved dyeing rate. The occurrence of some grooves on the fiber specimens was determined by scanning electron microscope and it was stated that these grooves might possibly provide a pathway for a faster dyeing rate [82].

Kan et al. (1998) treated the wool fiber with low-temperature plasma (LTP) using different non-polymerising gases and dyed with chrome dye. The rate of dyeing, the time of half-dyeing ($t_{1/2}$), the final dye uptake, the chromium exhaustion and the chromium fixation were studied. The results showed that LTP treatments can alter the dyeing properties to various degrees. The nature of the LTP gases plays an important role in affecting the behavior of chrome dyeing [83].

Kan et al. (1999), have studied the influence of the nature of gas (oxygen, nitrogen, and a 25% hydrogen/75% nitrogen gas mixture) in plasma treatments on the fiber-to-fiber friction, feltability, fabric shrinkage, surface structure, dyeability, alkali solubility, and surface chemical composition properties of wool substrates. After the low temperature plasma (LTP) treatment, those properties of the LTP-treated substrates changed, and the changes depended on the nature of the plasma gas used [84]. *Kan et al. (1999),* have also searched the surface characteristics of wool fibers treated with LTP with different gases, namely, oxygen, nitrogen and gas mixture (25% hydrogen / 75% nitrogen). Investigations showed that chemical composition of wool fiber surface varied differently with the different plasma gas used. The surface chemical composition of the different LTP-treated wool fibers was evaluated with different characterization methods, namely FTIR-ATR, XPS and saturated adsorption value [85].

Iriyama et al. (2002) treated the silk fabrics with O_2, N_2, and H_2 plasmas for deep dyeing and good color fastness to rubbing. C.I. Reactive Black 5 was used as a dye, and color was evaluated by total K/S. All plasma-treated silk fabrics showed weight loss, especially by O_2 plasma. Total K/S of dyed silk fabrics treated at 60 Pa of all plasmas was improved greatly. Total K/S increased with increasing plasma treatment time, weight loss of the fabrics in the treatment, and dye concentration in dyeing. They gained greater total K/S even dyed in 6% of dye concentration compared with untreated one dyed in 10%. Color fastness to wet rubbing of silk fabrics was not improved by plasma treatment. However, most of them were still within the level for commercial use [86].

Sun and Stylios (2004), have investigated the effects of LTP on pre-treatment and dyeing processes of cotton and wool. The contact angles, wicking properties, scourability, and dyeability of fabrics were affected by low-temperature plasma treatments. After treatment, the dye uptake rate of plasma treated wool has been shown to increase. It has been shown that O_2 plasma treatment increases the wetability of wool fabric and also the disulphide linkages in the exocuticle oxidize to form sulphonate groups which also enhances the wetability [87].

Binias et al. (2004) have investigated the effect of low temperature plasma on some properties (surface characteristic, water absorption capacity, capillarity, dyeability) of wool fibers. Selected properties of wool textiles were changed by the influence of low-temperature plasma on the fiber's surface layers, acting only on a very small thickness. The level of changes was limited by parameters of the low-temperature plasma. Lowering of the dyeing temperature was achieved [88].

Jocic et al. (2005), have investigated the influence of low-temperature plasma and biopolymer chitosan treatments on wool dyeability. Wool knitted fabrics were treated and characterized by whiteness and shrink-resistance measurements. Surface modification was assessed by contact-angle measurements of human hair fibers. It was stated that after plasma treatment the whiteness degree and hydrophility of fibers increased and fiber dyeability was improved [71].

Sun and Stylios (2005), have determined the mechanical and surface properties and handle of wool and cotton fabrics treated with LTP. This investigation showed that the mechanical properties of wool changed remarkably after oxygen plasma treatment. There were no significantly observed differences between plasma treated and un-treated fabrics after scouring and dyeing [89].

Masukuni and Norihiro (2006), studied the dyeing properties of Argon (Ar)-plasma treated wool using the six classes of dyestuffs, i.e., acid, acid metal complex, acid mordant, reactive, basic and disperse dyes. Ar-plasma treatment greatly improved the color yield and levelness, together with the decrease of tippy dyeing. A condition in the plasma treatment enhanced not only the color yield but also the anti-felting performance. The relationship between the improvement of dyeing properties by the plasma treatment and the chemical structure of the dyes was also examined. In the case of the acid dyes, the effect of plasma treatment on color yield was more significant for the milling type dyes with large molecular weight than the leveling type dye with low molecular weight. Furthermore, the hot water and rubbing fastness were improved by Ar-plasma treatment [90].

Kan and Yuen (2006) treated the wool fibers with oxygen plasma and then dyed these fibers with acid, chrome and reactive dye. For acid dyeing, the dyeing rate of the LTP-treated wool fiber was greatly increased, but the final dye uptake equilibrium did not show any significant change. For the chrome dyeing, the dyeing rate of the LTP-treated wool fiber was also increased, but the final dye uptake equilibrium was only increased to a small extent. For the reactive dyeing, the dyeing rate of the LTP-treated wool fiber was greatly increased, and the final dye uptake equilibrium was also increased significantly [91].

El-Zawahry et al. (2007) investigated the impact of plasma-treatment parameters on the surface morphology, physical-chemical, and dyeing properties of wool using anionic dyes. The LTP-treatment resulted in a dramatic improvement in fabric hydrophilicity and wettability, the removal of fiber surface material, and creation of new active sites along with improved initial dyeing rate. The nature of the plasma gas governed the final uptake percentage of the used acid dyes according to the following descending order: nitrogen plasma > nitrogen/oxygen (50/50) plasma >

oxygen plasma > argon plasma ≥ control. Prolonging the exposure time up to 20 minutes resulted in a gradual improvement in the extent of uptake [92].

Demir et al. (2008), were treated the knitted wool fabrics with atmospheric argon plasma, enzyme (protease), chitosan and a combination of these processes. The treated fabrics were evaluated in terms of their dyeability, color fastness and shrinkage properties, as well as bursting strength. The surface morphology was characterized by SEM images. In order to show the changes in wool surface after plasma treatment, XPS analysis was done. According to the experimental results it was stated that atmospheric plasma has an etching effect and increases the functionality of a wool surface [93].

Chvalinova and Wiener (2008) investigated the effects of atmospheric pressure plasma treatment on dyeability of wool fabric. Untreated and plasma treated wool materials were dyed with acid dye in weak acid solution (pH 6). Experiments showed invasion of surface layer of cuticle by plasma and it was observed that the plasma treated wool fabric for 100 seconds, absorbed double more dye than untreated wool fabric [94].

Naebe et al. (2010) treated the wool fabric with atmospheric-pressure plasma with helium gas for 30 seconds. X-ray photoelectron spectroscopy and time-of-flight secondary ion mass spectrometry confirmed removal of the covalently-bound fatty acid layer (F-layer) from the surface of the wool fibers, resulting in exposure of the underlying, hydrophilic protein material. Dye uptake experiments were carried out at 50°C to evaluate the effects of plasma on the rate of dye uptake by the fiber surface, as well as give an indication of the adsorption characteristics in the early stages of a typical dyeing cycle. The dyes used were typical, sulfonated wool dyes with a range of hydrophobic characteristics, as determined by their partitioning behavior between water and n-butanol. No significant effects of plasma on the rate of dye adsorption were observed with relatively hydrophobic dyes. In contrast, the relatively hydrophilic dyes were adsorbed more rapidly (and uniformly) by the plasma-treated fabric [95].

Demir (2010) treated the mohair fibers by air and argon plasma for modifying their some properties such as hydrophilicity, grease content, fiber to fiber friction, shrinkage, dyeing, and color fastness. The results showed that the atmospheric plasma has an etching effect and increases the functionality of a fiber surface. The hydrophilicity, dyeability, fiber friction coefficient, and shrinkage properties of mohair fibers were improved by atmospheric plasma treatment [96].

Atav and Yurdakul (2011) investigated the use of plasma treatment for the modification of fiber surfaces to achieve dyeing of mohair fibers at lower temperatures without decreasing dye uptake. The study was carried out by using different gases under various powers and times. The effect was assessed in terms of color. Test samples were also evaluated using scanning electron microscopy (SEM). The optimum conditions of plasma treatment for improving mohair fiber dyeability, is treatments carried out by using Ar gas at 140 W for 60''. According to the experimental results it can be concluded that plasma treated mohair fibers can be dyed at lower temperatures (90°C) shorter times (1 h instead of 1.5 h) with reactive dyes without decreasing color yield. Dyeing kinetics was also searched in the study and it was demonstrated that the rate constant and the standard affinity of plasma treated sample was increased [97].

6. GAMMA IRRADIATION TECHNOLOGY

Gamma radiation, also known as gamma rays or hyphenated as gamma-rays and denoted as γ, is electromagnetic radiation of high frequency and therefore energy. Gamma rays typically have frequencies above 10 exahertz (or $>10^{19}$ Hz), and therefore have energies above 100 keV and wavelengths less than 10 picometers (less than the diameter of an atom) [98]. Gamma rays are identical in nature to other electromagnetic radiations such as light or microwaves but are of much higher energy. Examples of gamma emitters are cobalt-60, zinc-65, cesium-137, and radium-226. Like all forms of electromagnetic radiation, gamma rays have no mass or charge and interact less intensively with matter than ionizing particles [99].

Studies on the interaction of high energy radiation with polymers have attracted the attention of many researchers. This is due to the fact that high-energy radiation can induce both chain scission and/or crosslinking [100]. The efficiency of these two types of reactions depends mainly on polymer structure and irradiation atmosphere. However, dose rate, types of radiation source and temperature during irradiation can influence the reaction rates [101]. For some applications radiation degradation can be controlled and devoted to achieve a specific property [100]. Degradation, in broad terms, usually involves chemical modification of the polymer by its environment; modification that is often (but not always) detrimental to the performance of the polymeric material. Although the chemical change of a polymer is frequently destructive, for some applications degradation can be controlled and encouraged to achieve a specific property. In this regard, different vinyl monomers have been grafted onto gamma irradiated wool fabric to improve some favorable properties such as dyeability, and moisture regain. These studies have been based on the formation of stable peroxides on wool, upon irradiation, which are thermally decomposed to initiate polymerization [102].

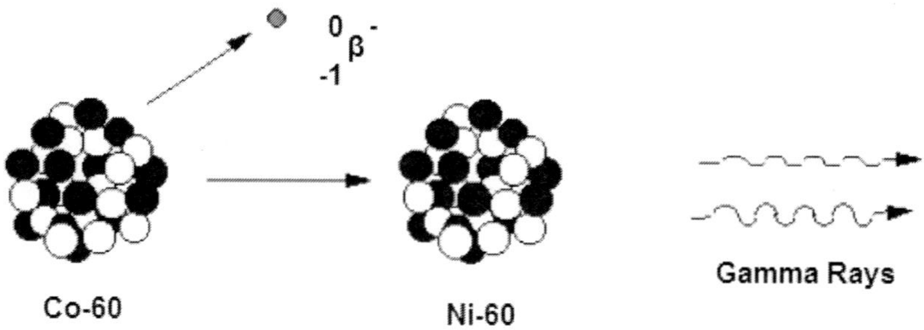

Figure 11. Gamma ray radiation [**99**]

Gamma rays are ionizing radiations that interact with the material by colliding with the electrons in the shells of atoms. They lose their energy slowly in material being able to travel through significant distances before stopping. The free radicals formed are extremely reactive, and they will combine with the material in their vicinity. The irradiated modified fabrics can allow: more dye or pigment to be fixed, producing deeper shades and more rapid fixation of dyes at low temperature [32]. In literature it is stated that two kinds of effects might occur in parallel in wool during the irradiation. The first effect as manifests as an evident decrease in dye accessibility at lower doses may not be altogether independent of crosslinking. On the other hand, the remarkable increase in the uptake at higher doses seems to be associated with strong structural damage of fibers. It is interesting to note that the increase in accessibility to dyes of the highly irradiated fibers is so great that the bilateral structure is hardly visualized by the partial staining. Thus the cross-sections of fibers irradiated with a dose of 10^{8}roentgens are stained uniformly in dark tone even under the condition which does not give rise to the staining of unirradiated fibers

[103]. In literature there are limited studies related to the effect of gamma irradiation treatment on dyeability of proteinous fibers. Some of them are summarized below.

Horio et al. (1962) investigated the effect of irradiation on dyeing property of wool fiber. It was found that the dyeing property of wool fiber was greatly affected by irradiation even at low doses where any changes in mechanical properties were not noticeable. The rate of dye absorption was strongly depressed by irradiation with Co-60 gamma radiation of low doses from 10^3 to 10^5 roentgens. Two dyestuffs, C.I. Acid Red 44 and C. I. Acid Green 28 were used. The dye absorption was strikingly suppressed at the range of doses from 10^3 to 10^5 roentgens, but fibers regain dye accessibility at higher doses [103].

Beevers and McLaren (1974) have been found that small doses of gamma radiation (0.5-10 Mrad) produce marked effects on some physical properties of wool. The results indicate that even small doses of gamma radiation break sufficient covalent bonds to make the crosslinked peptide chain structure more susceptible to the action of swelling and disordering agents. These small radiation-induced changes can be expected to affect properties of wool significantly in absorption and penetration processes, such as those involved in dyeing, chemical modification, and grafting treatments of wool [104].

Millington (2000) investigated the effects of γ-radiation (^{60}Co) on some chemical and physical properties of wool keratin and compared and contrasted with the effects of ultraviolet radiation in the UVC (200-280 nm) region. The effect of UVC doses up to 25 J/cm^2 on fabric strength was found to be small (5%), whereas γ-irradiated wool experienced strength reductions of 15% at doses over 100 kGy. Color changes following UVC and γ-irradiation were quite different: UVC wool was initially green changing to yellow under ambient conditions, γ-treated wool became pink-red at doses 25-250 kGy, and yellow at higher doses. The chromophores produced by UVC were easily removed by oxidative bleaching with hydrogen peroxide, whereas γ-treated wool remained yellow even at relatively low doses (25–50 kGy). This has implications for the use of γ-radiation as a means of sterilising wool for compliance with quarantine regulations. The effects of the two forms of radiation on the natural fluorescence of wool, permanent setting, printing properties and the epicuticle layer were also described [105].

7. LASER TECHNOLOGY

In the last decade, considerable effort has been made in developing surface treatments such as UV irradiation, plasma, electron beam and ion beam to modify the properties of textile materials. Laser modification on material surface is one of the most studied technologies [106]. A laser is a device that emits light (electromagnetic radiation) through a process of optical amplification based on the stimulated emission of photons. The term "laser" originated as an acronym for "Light Amplification by Stimulated Emission of Radiation" [107]. Laser processing as a new processing method, with its processing of accurate, fast, easy, automatization, in leather, textile and garment industry increasingly widely used [108].

Laser technology has been widely used in surface modification of polymers [109]. Since the late 90s, different types of commercial lasers are available for surface modification of materials [110]. While most of the efforts in developing surface treatments have been made using UV laser, infra-red lasers, like CO_2 appear to be less concerning [109]. Adequate power levels for a specific application are very important in surface modification processes, because an excessive amount of energy can damage the polymers. Infra-red lasers like CO_2 are the most powerful lasers; with no suitable power level, severe thermal damage can be resulted. However, this short coming can be overcome by the use of pulsed-mode CO_2 lasers, which are easier to control than lasers operating in the continuous wave mode [110]. Excimer lasers, which are a form of ultraviolet lasers [111], are a special sort of gas laser powered by an electric discharge in which the lasing medium is an excimer, or more precisely an exciplex in existing designs. These are molecules which can only exist with one atom in an excited electronic state. Once the molecule transfers its excitation energy to a photon, therefore, its atoms are no longer bound to each other and the molecule disintegrates. This drastically reduces the population of the lower energy state thus greatly facilitating a population inversion. Excimers currently used are all noble gas compounds; noble gasses are chemically inert and can only form compounds while in an excited state. Excimer lasers typically operate at ultraviolet wavelengths [107].

It has been shown that materials like polymers, woods, metals, semiconductors, dielectrics and quartz modified by laser irradiation often exhibit physical and chemical changes in the material's surface [106]. Physical modifications occur in the form of a certain, regular surface structure of the irradiated sites. The high energy input of the excimer radiation into the polymer might also give rise to chemical changes of the surface [112]. In the case of polymers, some well-oriented structure of grooves or ripple structures with dimensions in the range of micrometer are developed on surface with irradiation fluence above the so-called ablation threshold. The laser irradiation of highly absorbing polymers can generate characteristic modifications of the surface morphology. The physical and chemical properties are also affected after laser irradiation. Hence, it is reasonable to believe that such surface modification of a polymer may have an important impact on its textile properties [106].

The possible applications of laser technology in the textile industry include removal of indigo dye of denim, heating threads, creating patterns on textiles to change their dyeability, producing surface roughness, welding, cutting textile webs. Laser irradiation on polymer surface is used to generate a modified surface morphology. The smooth surface of polymers is modified by this technique to a regular, roll-like structure that can cause adhesion of particles and coating, wetting properties and optical appearance [113].

Laser technology can also be used for improving dyeability since it is well known that the UV output from excimer lasers can modify the surface of synthetic fibers. But the use of laser energy in textile treatment is still uncommon and is generally limited with denim garment finishing [114]. Although there are many studies in the literature related to the investigation of the effect of laser treatment on dyeability of many fiber types, there is still no study carried out on wool. But, by taking into consideration that the chemical structure and dyeing property of proteinous fibers is similar to polyamide, in the light of the studies carried out on polyamide [114, 115], it can be said that dyeability of proteinous fibers with anionic dyes such as acid and reactive dyes may increase due to the decrease in the crystallinity and increase in free amino groups of the fiber.

8. MICROWAVE TECHNOLOGY

Microwaves (MW) which have broad frequency spectrum are electromagnetic waves that are used in radio, TV and radar technology [116]. MWs are radio waves with wavelengths ranging from as long as one meter to as short as one millimeter, or equivalently, with frequencies between 300 MHz (0.3 GHz) and 300 GHz [117].

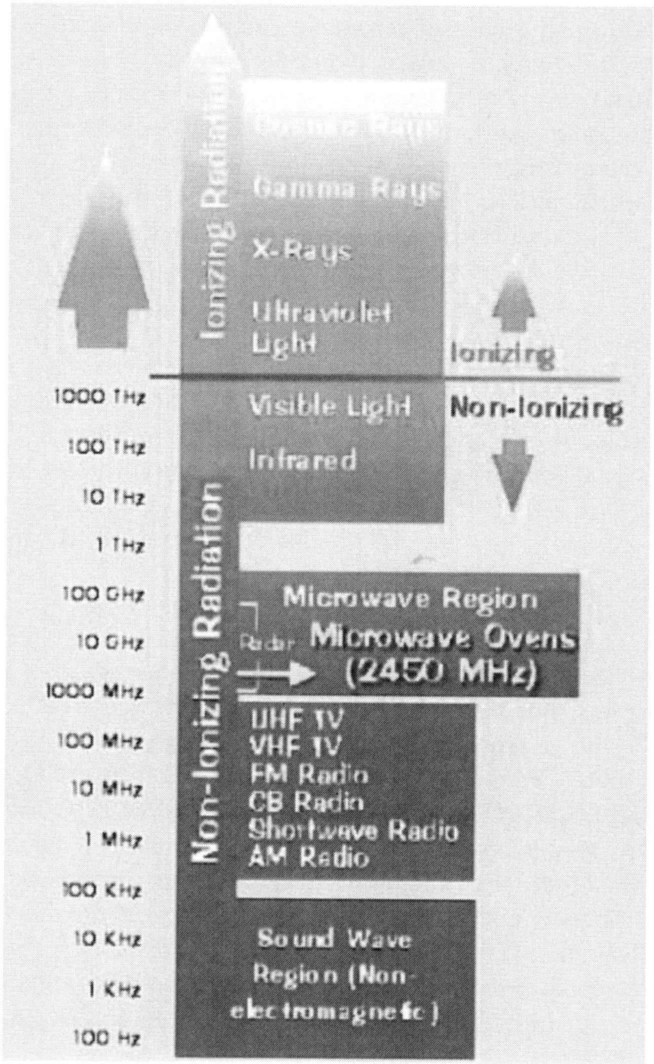

Figure 12. Frequency spectrum [**118**]

The term "microwave" denotes the techniques and concepts used as well as a range of frequencies. Microwaves travel in matter in the same manner as light waves: they are reflected by metals, absorbed by some dielectric materials and transmitted without significant losses through other materials. For example, water, carbon, foods with a high water, some organic solvents are good microwave absorbers whereas ceramics, quartz glass and most thermoplastic materials absorb microwaves slightly [119].

Electromagnetic waves can be absorbed and be left as energy units called photon. The energy carried by photon is depended on the wavelength and the frequency of radiation. Energy of MW photons is 0.125 kJ/mol. This value is very low considering the necessary energy for chemical bonds. Therefore MW rays can not affect the molecular structure of the material directly and change the electronic structures of atoms [116].

Microwave-promoted organic reactions are known as environmentally benign methods that can accelerate a great number of chemical processes. In particular, the reaction time and energy input are supposed to be mostly reduced in the reactions that are run for a long time at high temperatures under conventional conditions [120]. Reactions conducted through microwaves are cleaner and more environmentally friendly than conventional heating methods [121].

The use of microwave radiation as a method of heating is over five decades old. Microwave technology originated in 1946, when Dr. Percy Le Baron Spencer, while conducting laboratory tests for a new vacuum tube called a magnetron, accidentally discovered that a candy bar in his pocket melted on exposure to microwave radiation. Dr. Spencer developed the idea further and established that microwaves could be used as a method of heating. Subsequently, he designed the first microwave oven for domestic use in 1947. Since then, the development of microwave radiation as a source of heating has been very gradual [121].

Microwave heating occurs on a molecular level as opposed to relying on convection currents and thermal conductivity when using conventional heating methods. This offers an explanation as to why microwave reactions are so much faster [122]. The fundamental mechanism of microwave heating involves agitation of polar molecules or ions that oscillate under the effect of an oscillating electric or magnetic field. In the presence of an oscillating field, particles try to orient themselves or be in phase with the field. However, the motion of these particles is restricted by resisting forces (inter-particle interaction and electric resistance), which restrict the motion of particles and generate random motion, producing heat [121]. Microwave (MW) heat systems consists of three main units; magnetron, waveguide and applicator. Magnetron is used as a microwave energy source in industrial and domestic type of microwave ovens. One of the oscillator tube, magnetron consists of two main parts as anode - cathode, and it converts the continuous current - electrical energy to MW energy. Circulator transmits approximately all of the waves that are sent from magnetron and shunts transmitted waves to water burden. Thus magnetron is protected. Electromagnetic waves are transmitting to the applicator by waveguides. Applicators are parts of the matter MW applied on. MW energy produced in generator is affected directly on the material in applicators in MW heating systems. Type of applicators used in practice can be divided into three groups as multi-mode (using 80% of the industrial systems), single-mode and near field MW applicators [116].

Since the beginning of the twentieth century MW technology has made significant contributions to scientific and technological developments. Also due to its initial intend to be used in telecommunications, very important progresses have been made in this area. Nevertheless from the second half of twentieth century, MW energy is finding increased number of application area in other industrial processes and these applications are surpassing telecommunication applications [116].

In textile processing it is necessary to apply heat as in dye fixation, heat setting or drying the product. Heat can be transferred to the material by radiation, conduction and convection. These three ways of transferring can be used either separately or in combination. The saving of time and energy is of immediate interest to the textile industry. The introduction of new techniques which will allow less energy to be used: is a highly important area of activity to consider. The textile industry has investigated many uses for microwave energy such as heating, drying, dye fixation, printing and curing of resin finished fabrics. In 1966, Ciba-Geigy obtained one of the earliest patents for using microwave heating in dyeing and printing fibrous material with reactive dyes. Since then many authors have investigated the feasibility of using microwaves for a variety of dyeing and finishing processes [123]. Although many studies have focused on investigating the feasibility of using microwaves to dye polyester fibers with disperse dyes, researches related to the use of microwave heating in dyeing of proteinous fibers are very limited.

Delaney and Seltzer (1972) used microwave heating for fixation of pad-dyeings on wool and they demonstrated the feasibility of applying certain reactive dyes to wool in fixation times of 30-60 s [124].

Zhao and He (2011) treated the wool fabric with microwave irradiation at different conditions and then studied for its physical and chemical properties using a variety of techniques, such as Fourier transform infrared spectroscopy, X-ray diffraction, and scanning electron. It was found that microwave irradiation of wool fabric significantly improved its dyeability. It was stated that this could be due to the change in wool surface morphological structure under microwave irradiation which implied that the barrier effect in wool dyeing was diminished. Although the breaking strength of the treated wool fabrics also improved with microwave irradiation, the chemical structure and crystallinity did not show any significant change [125].

In literature it was stated that dyeing time of mohair could be drastically reduced from the conventional 90 minutes to 35 minutes using the radio frequency technique, only 5 minutes of the 35 minutes representing actual exposure to the radio frequency field. An estimated saving of some 80% in dyeing energy costs could be achieved. Furthermore, dye fixation improved slightly from 93% to 96%.*Turpie* quoted unpublished data in which radio-frequency dyeing of tops produced better luster and enabled higher maximum spinning speeds than did conventional dyeing of mohair. *Smith,* claimed advantages of radio-frequency dyeing and bleaching, - for example, reduced chemical damage because of a shorter exposure to 100°C; and reduced energy, water and effluent costs [126].

9. E-BEAM IRRADIATION TECHNOLOGY

Radiation processing has been increasingly applied in industries to improve the quality of products, efficiency, energy saving and to manufacture products with unique properties [127]. An irradiation can induce chemical reactions at any temperature in the solid, liquid and gas phase without any catalyst; it is a safe method that could protect the environment against pollution; radiation process could reduce curing time and energy saving; it could treat a large and thick three-dimensional fabrics that need not consider of the shape of the samples [128].

Physical techniques for activating fiber molecules in the absence of solvent for producing functional textiles are becoming increasingly attractive also from an ecological viewpoint. Among them, electron beam processing is particularly interesting as it offers the possibility to treat the materials without solvent, at normal temperature and pressure [129]. Whilst the energy of the electrons in gas discharge plasmas is typically in the range of 1-10 electron volt (eV), electron-beam (E-beam) accelerators generate electrons with a much higher energy, generally 300 keV to 12 MeV. These electrons may be used to modify polymer materials through direct electron-to-electron interactions. These interactions can create active species such as radicals, so there are different possible outcomes from the electron-beam irradiation of polymer materials, on the basis of the chosen operating conditions [130].

The formation of active sites on the polymer backbone can be carried out by several methods such as plasma treatment, ultraviolet (UV) light radiation, decomposition of chemical initiator and high-energy radiation [131]. At present, the most common radiation types in industrial use are gamma and e-beam. E-beam machines play a significant role in the processing of polymeric materials; a number of different machine designs and different energies are available. Industrial e-beam accelerators with energies in the 150-300 keV range are in use in applications where low penetration is needed, such as curing of surface coatings. Accelerators operating in the 1.5 MeV range are used where more penetration is needed. E-beam machines have high-dose rate and therefore short processing times. While they have limited penetration compared with gamma, they conversely have good utilization of energy because it can all be absorbed by the sample irradiated [132].

New radiation processing applications involving ion-beam treatment of polymers offer exciting prospects for future commercialization, and are under active investigation in many research laboratories. Much of the work has involved the irradiation of non-polar polyolefins (PE, PS, PP, etc.) as a means of inducing polar groups at surfaces, in order to enhance such properties as printability, wetability and exc. [132]. Studies related to the e-beam irradiation of proteinous fibers are limited.

Fatarella et al.(2010) wanted to clarify whether fiber surface treatments such as plasma and e-beam could affect the effectiveness of a TGase treatment by improving the accessibility of target amino acids in wool to the enzyme or improving the penetration of the enzyme into the wool fiber cortex, thus accessing more sites for reaction. Plasma treatments with different non-polymerizing gases (oxygen, air and nitrogen) and e-beam irradiation in air or nitrogen atmosphere were assessed as possible pretreatments to non-proteolytic enzymatic processes (such as TGase) to improve the accessibility of target groups in the wool proteins to the enzymes. Only limited inhibition and/or inactivation of the transglutaminase enzyme was found after treatment with plasma and e-beam, suggesting such treatments could be used as a preparative treatment prior to the application of the enzyme. In contrast, by increasing the energy of the electrons in e-beam treatments no significant superficial modifications were observed. In fact, they promoted the cleavage of high-energy bond, such as S-S linkage, by enhancing depolymerization reaction [130].

10. ION IMPLANTATION TECHNOLOGY

It has been approximately 50 years since researchers first began exposing polymeric materials to ionizing radiation, and reporting the occurrence of cross-linking and other useful effects. Innovation in this field has by no means ended; important new products made possible through radiation technology continue to enter the marketplace, and exciting new innovations in the application of radiation to macromolecular materials are under exploration at research institutions around the world [132]. Ionizing radiation may modify physical, chemical and biological properties of materials [133]. Some of the surface characteristics being successfully manipulated by radiation grafting include: chemical resistance, wetability, biocompatibility, dyeability of fabrics, and antistatic properties [132].

Ion implantation is an innovative production technique with which the surface properties of inert materials can be changed easily. It shows distinct advantages because it is environmentally friendly. Ion implantation can be used to induce both surface modifications and bulk property enhancements of textile materials, resulting in improvements to textile products ranging from conventional fabrics to advanced composites. Ion implantation was first done by Rutherford in 1906, when he bombarded aluminum foil with helium ions. Ion implantation has been applied to metals, ceramics, plastics, and polymers [134].

Even though ion implantation is relatively complex in terms of the equipment required, it is a relatively simple process. Ion implantation consists of basically two steps: form plasma of the desired material, and either extract the positive ions from the plasma and accelerate them toward the target, or find a means of making the surface to be implanted the negative electrode of a high voltage system [135]. There are three methods commonly used for ion implantation. They differ in the way in which they either form the plasma or make the surface to be implanted the negative electrode. These methods are mass-analyzed ion implantation, direct ion implantation and plasma source ion implantation [134].

In *mass-analyzed ion implantation*, the plasma that is formed in the ion source is not pure; it contains materials that one does not wish to implant.

Thus, these contaminants must be separated from the plasma. To perform this separation, the plasma source is placed at a high voltage and the part to be implanted is placed at ground. This produces a situation where the target is at a negative potential with respect to the plasma source. A negative electrode then extracts the ions from the source. The ions are then accelerated by a high voltage source to the target. Between the ion source and the target there is a large magnet, with magnetic field perpendicular to the direction of ion motion. Ions passing through this magnetic field are bent by the magnetic field. The amount of bending depends on the ion material being implanted and the strength of the magnet [135].

Direct ion implantation eliminates the need for the current limiting magnet found in mass-analyzed ion implantation by using an ion source that produces a plasma and ion beam of just the desired material. In direct ion implantation, the plasma is formed in the ion source and the ions extracted at high energies in a wide beam, passing through a valve directly into the end station, where the ion implant parts within the target area. In such a case, the beam current density can be high (10-50 mA), costs are greatly reduced, and relatively high throughput processing is possible [135].

It is difficult in practice to treat uniformly three-dimensional objects such as sockets in artificial hip joints, without using sophisticated manipulating devices. Recently, however, an innovative hybrid technology, *plasma source ion implantation* has been developed, which can, to a large extent, address the problems with conventional ion implantation of components with complex shapes [136]. In plasma source ion implantation (sometimes referred to as plasma ion immersion), the plasma source floods the chamber of the end station with plasma. Ions are extracted from the plasma and directed to the surface of the part being ion implanted by biasing the part to very high negative voltages using a pulsed, negative high voltage power supply [135].

One of the most recent ion implantation techniques is *metal vapor vacuum arc (MEVVA)*, a type of direct ion implantation. In the mid-1980s, Brown et al. at Lawrence Berkeley Laboratory developed a new type of metal ion source, namely the MEVVA ion source. The MEVVA source makes use of the principle of vacuum arc discharge between the cathode and the anode to create dense plasma from which an intense beam of metal ions of the cathode material is extracted. This metal ion source operates in a pulse mode [134].

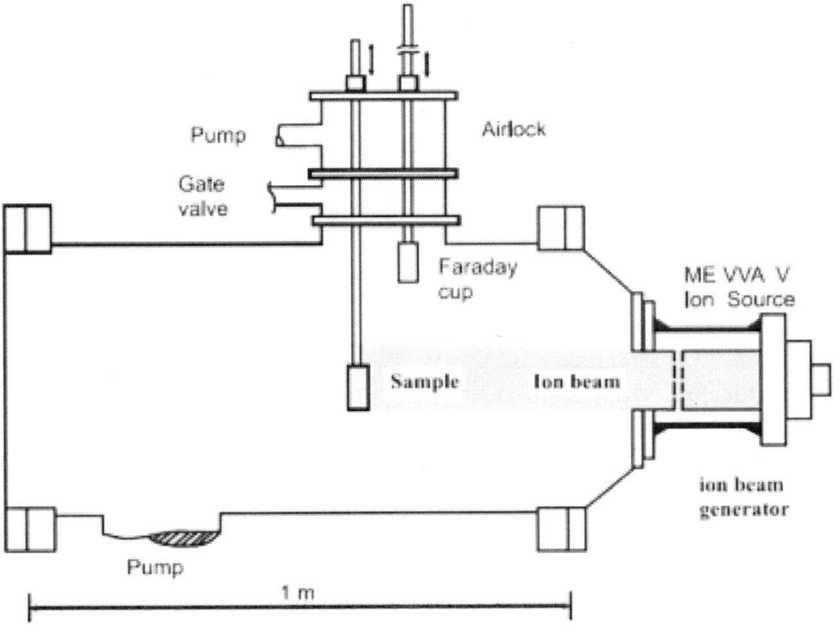

Figure 13. MEVVA ion implanter [**134**]

Many techniques have been applied to produce the great potential of ion beam modification technology. It desired surface modifications, ranging from conventional flame treatments, ''wet'' chemical treatments, and electrical treatments (such as corona discharge), to modern plasma treatments and particle beam irradiation (electrons, ions, neutrons and photons) techniques. Among them, particle beam techniques are particularly attractive owing to their flexibility, effectiveness, and environmentally friendly nature compared with conventional techniques. Also, in the domain of particle beam techniques, the ion beam has proven more effective in modifying polymer surfaces than UV-light, c-ray, X-ray and electron beams. This is because energetic ions have a higher cross-section for ionization and larger linear energy transfer (LET, eV nm^{-1}) than these conventional radiation types of comparable energy owing to their deeper range [136].

Numerous papers have appeared in recent years describing surface treatment using ions. Surface modification with ions typically involves fluencies of ~ 10^9 to 10^{14} ions/cm^2 or in some cases, 10^{15}ions/cm^2; higher fluencies may result in destruction of polymer, generally through carbonization. Many different ions have been employed for irradiating polymers, ranging from hydrogen and helium ions up to ions of heavy elements such as gold or uranium [132].

11. SUPERCRITICAL CARBONDIOXIDE TECHNOLOGY

The enhanced solvent properties provided by supercritical fluids (SCFs) are by no means new. In fact, the various phenomena attributed to SCFs have been known for hundreds of years [137]. However, textile dyeing, using supercritical carbon dioxide as dye solvent, has been developed for the last two decades since it potentially overcomes all the economical and ecological disadvantages derived from the conventional water dyeing process [138]. Supercritical fluids are characterized by a very high solute diffusivity and low viscosity. Consequently, in supercritical fluids all the transport processes are much more rapid. Dyeing in supercritical carbondioxide, for example, entirely avoids the use of water, and consequently there is no possibility of pollution. No auxiliary agents are used and residual dye can be recovered in a reusable form [139].

A supercritical fluid can be defined as a substance above its critical temperature and pressure. Under this condition the fluid has unique properties, in that it does not condense or evaporate to form a liquid or a gas [140]. Supercritical fluids exhibit gas-like viscosities and diffusivities and liquid-like densities [120]. Although a number of substances are useful as supercritical fluids, carbon dioxide has been the most widely used [140], as the critical temperature and pressure are easier to achieve than that of other substances [141]. Supercritical CO_2 is a dyeing medium which is a potential alternative to water as it is inherently nontoxic, inexpensive, and nonflammable, it can be recycled, it has easily accessible critical conditions [142]. The critical point for carbon dioxide occurs at a pressure of 73.8 bar and a temperature of 31.1°C. $ScCO_2$ represent a potentially unique media for either transporting chemical into or out of a polymeric substrate, because of their thermo-physical and transport properties. The phase diagram of carbon dioxide shown in Fig. 14 represents the interfaces between phases; at the triple point all three phases may coexist. Above the triple point, an increase in temperature drives liquid into the vapor phase, while an increase in pressure drives vapor back to liquid [120]. Above the critical point of carbon dioxide, it retains the free mobility of the gaseous state, but with the increasing pressure its density will increase towards that of a liquid. Solvating power is proportional to density, whilst viscosity remains comparable with that of a normal gas, so the "fluid" has remarkable penetration properties [143].

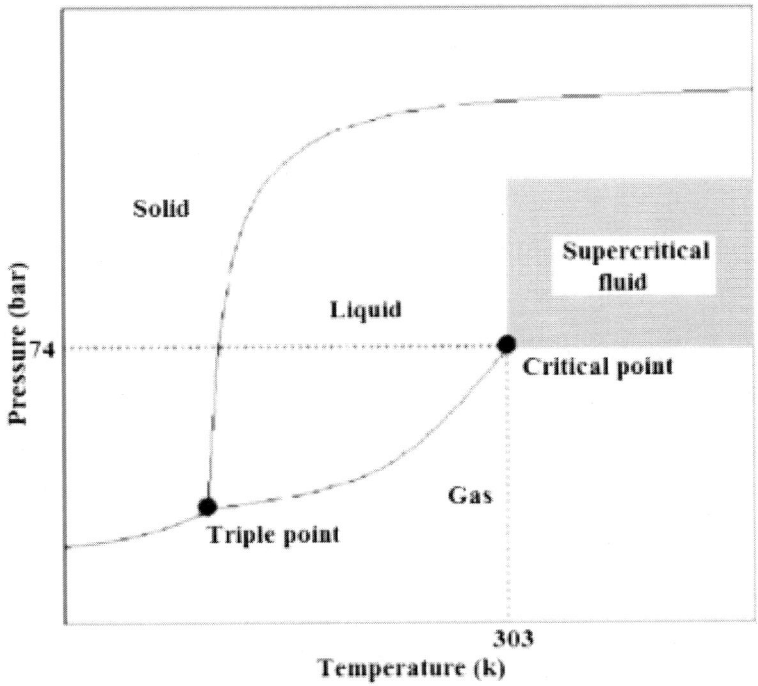

Figure 14. Phase diagram of carbon dioxide [**120**]

In the dyeing field, Schollmeyer and co-workers, among others, have demonstrated that both synthetic and natural fibres can be dyed with disperse and reactive disperse dyes in supercritical carbondioxide. However, the dyeing of natural fibres with water soluble dyes in supercritical carbondioxide has not yet been successful, since dyes such as reactive, acid and basic dyes have little solubility in this medium due to their high polarity [144]. One approach to this problem was undertaken by Gebert et al. who examined wool and cotton fibers after attempting to open the fiber surface with a swelling auxiliary so that dye molecules could be readily trapped in the fiber [145].

In natural textiles, the dye molecules can be fixed by either physical (e.g. Van der Waals) or chemical (e.g. covalent) bonds. Since the dyes used in a ScCO2-dyeing process are non-polar and natural fibres are polar, the affinity between dyes and textiles is low so physical bonds are weak.

Therefore, a dyeing process must be developed for dyeing natural textiles in ScCO2 with reactive dyes that create covalent dye-textile bonds. So far, several reactive dyes known from conventional dyeing in water have been investigated in ScCO2:

- vinylsulphone dyes have been successfully used for silk and wool,
- 2-bromoacrylamide dyes have been successful in dyeing wool and cotton,
- dichlorotriazine dyes have been tested on silk and cotton but showed insufficient fixation [146].

Guzel and Akgerman (2000) dyed the wool fibers with three mordant dyes dissolved in supercritical carbon dioxide. Wool fibers were mordanted with five different metal ions (Cr(III), Al(III), Fe(II), Cu(II) and Sn(II)) using conventional techniques and dyed at 333-353°K temperature and at 150-230 atm pressure. According to the experimental results it was found that dyed materials had excellent wash fastness properties [147].

For water-soluble dyes, attempts were made to dye natural fibers using reverse micelle technique (Fig. 15) in which ionic dye, solubilized in the water-pool, passes into the fiber together with a small amount of water immediately after contact with it. Satisfactory results were obtained for proteinous fibers [120].

Sawada et al. (2002) has developed a reverse micellar system in supercritical carbon dioxide as a dyeing medium. Water-soluble dyes such as reactive dyes and acid dyes could be sufficiently solubilised in the interior of a specially constituted reverse micelle. Protein fabrics, silk and wool, were satisfactorily dyed even in deep shades with conventional acid dyes without any special pretreatment. Compared to previously proposed supercritical dyeing methods, dyeing of fabrics with this system could be performed at low temperatures and pressures in a short time [148].

Jun et al. (2004) investigated the dyeing of wool fabrics with conventional acid dyes in a supercritical CO_2 using a reverse micellar system. A reverse micelle composed of perfluoro 2,5,8,11-tetramethyl-3,6,9,12-tetraoxapentadecanoic acid ammonium salt/CO_2/water had a high potential to solubilize conventional acid dyes and to dye wool fabrics in this system. It was found that dyeability of the acid dye on wool in this system take no influence of the density of CO_2. On the other hand, variation of dyeing temperature resulted in the remarkable differences of the dyeability of the acid dye on wool even though the solubility of dye in the system was not varied by the variation of temperature [149]

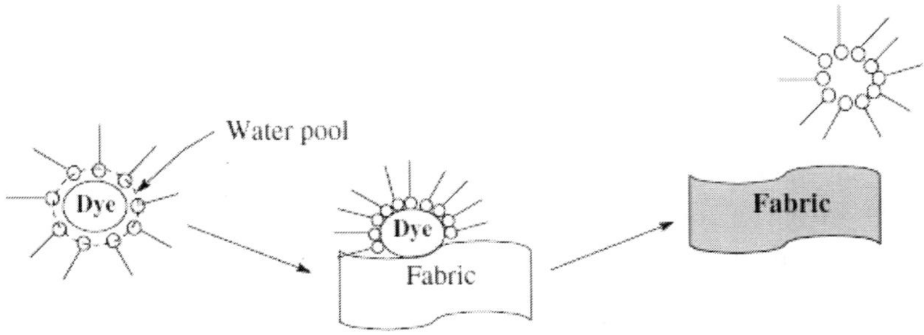

Figure 15. Dyeing of natural fibers using reverse micelle [**120**]

Jun et al. (2005) investigated the phase behavior of a cationic perfluoropolyether surfactant in supercritical carbon dioxide and its ability to solubilise ionic dyes. This cationic surfactant was found to dissolve satisfactorily in supercritical carbon dioxide and was able to form micellar aggregates incorporating a small amount of water in their interior. Conventional anionic dyes such as acid and reactive dyes were solubilised satisfactorily in the cationic surfactant/water/supercritical carbon dioxide system. The surfactant was more effective in this respect than its anionic analogue [144].

Schmidt et al. (2007) dyed various fibers, among which wool and silk also exist, with C.I. Disperse Yellow 23 modified with 2-bromoacrylic acid and 1,3,5-trichloro-2,4,6-triazine as reactive groups in supercritical carbon dioxide. It was found that on wool and silk, the color depth was higher than on cotton and the wash, rub and light fastness of all dyeings was between 4 and 5 [150].

12. CONCLUSIONS

Textile dyeing is the most remarkable process among the wet treatments in textile industry in terms of energy and water consumption and effluent load. In last two decades increased laws related to the environment and competitive market conditions required some new processes to be found in textile dyeing field. This situation increased the interest of the usage of new technologies such as ultrasound, ultraviolet, ozone, plasma, gamma irradiation, laser, microwave, e-beam irradiation, ion implantation, and supercritical carbondioxide in textile industry. These new technologies provide not only decrease in time, energy, and chemical consumption, but also decrease in effluent load. So that all of these new technologies considered to be very interesting future oriented processes because of being environmentally friendly. Although it was proven with many researches that most of these technologies are successful at laboratory scale, there is still need to integrate them into industrial applications. There is no doubt that in future these new technologies will find wide range of applications when their disadvantages (to be expensive, not possible to be used for all fiber types and exc.) will be eliminated.

REFERENCES

1. Lewis, D.M., 1992, Wool Dyeing, Society of Dyers and Colorists
2. Jocic, D., Vilchez, S., Topalovic, T., Molina, R., Navarro, A., Jovancic, P., Julia, M.R. and Erra, P., 2005, Effect of Low-Temperature Plasma and Chitosan Treatment on Wool Dyeing with Acid Red 27, Journal of Applied Polymer Science, 97 (6), 2204–2214p.
3. Molina, R., Erra, P., Julia, L., and Bertran, E., 2003, Free Radical Formation in Wool Fibers Treated by Low Temperature Plasma, Textile Research Journal, 73, 955-959p.
4. Onar, N. and Sarıışık M., January-March 2005, Use of Enzymes and Chitosan Biopolymer in Wool Dyeing, Fibers & Textiles in Eastern Europe, 13 (1), 54-59p.
5. Cardamone, J. M. and Damert, W.C., 2006, Low-Temperature Dyeing of Wool Processed for Shrinkage Control, Textile Research Journal, 76 (1), 78–85p.
6. Kongkachuichay, P., Shitangkoon, A. and Chinwongamorn, N., 2002, Studies on Dyeing of Silk Yarn with Lac Dye: Effects of Mordants and Dyeing Conditions, Science Asia, 28, 161-166p.
7. Luo, J. and Cao, J., 1997, Low-Temperature Dyeing of Real Silk Fabrics with Liquid Sulphur Dyes, JSDC, 113, 67-69p.

8. Leksophee, T., Supansomboon, S. and Sombatsompop, N., 2004, Effects of Crosslinking Agents, Dyeing Temperature, and pH on Mechanical Performance and Whiteness of Silk Fabric, Journal of Applied Polymer Science, 91, 1000-1007p.

9. Tsukada, M., Katoh, H., Wilson, D., Shin, B., Arai, T., Murakami, R. and Freddi, G., 2002, Production of Antimicrobially Active Silk Proteins by Use of Metal-Containing Dyestuffs, Journal of Applied Polymer Science, 86, 1181-1188p.

10. Shao, J., Liua, J. and Carr, C.M., 2001, Investigation into The Synergistic Effect Between UV/Ozone Exposure and Peroxide Pad-Batch Bleaching on The Printability of Wool, Coloration Technology, 117 (5), 270-275p.

11. Kamel, M.M., El-Shistawy, R.M., Hana, H.L. and Ahmed, N.S.E., 2003, Ultrasonic-Assisted Dyeing: I. Nylon Dyeability with Reactive Dyes, Polymer International, 52 (3), 373-380p.

12. Akalın, M., Merdan, N., Kocak, D. and Usta, I, 2004, Effects of Ultrasonic Energy on The Wash Fastness of Reactive Dyes, Ultrasonics, 42, 161-164p.

13. Vajnhandl, S. and Le Marechal, A.M., 2005, Ultrasound in Textile Dyeing and The Decoloration/Mineralization of Textile Dyes, Dyes and Pigments, 65 (2), 89-101p.

14. Boz, S., 2008, Ultrasonik Enerjinin Konfeksiyon Sanayiinde Kullanımının İncelenmesi, Master Thesis Ege University Textile Engineering Department

15. Moholkar, V.S., Nierstrasz, V.A. and Warmoeskerken, M.M.C.G., 2003, Intensification of Mass Transfer in Wet Textile Processes by Power Ultrasound, AUTEX Research Journal, 3 (3), 129-138p.

16. Jung, C., Budesa, B., Fässler, F., Uehlinger, R., Müller, T. and Wild, M., Effect of Cavitation in Ultrasound Assisted Cleaning of Medical Devices, http://www.kks-ultraschall.ch/assets/files/news/effect-of-caviation-ultrasound.pdf

17. Kamel, M.M., El-Shishtawy, R.M., Yussef, B.M. and Mashaly, H., 2005, Ultrasonic assisted dyeing III. Dyeing of wool with lac as a natural dye, Dyes and Pigments, 65 (2), 103-110p.

18. Moholkar, V.S., Rekveld, S. and Warmoeskerken, M.M.C.G., 2000, Modeling of The Acoustic Pressure Fields and The Distribution of The Cavitation Phenomena in A Dual Frequency Sonic Processor, Ultrasonics, 38, 666-670p.

19. Santos, H.M., Lodeiro, C., Luis, J.L. and Martinez, C., The Power of Ultrasound, http://www.wiley-vch.de/books/sample/3527319344_c01.pdf

20. Vouters, M., Rumeau, P., Tierce, P. and Costes, S., 2004, Ultrasounds: An Industrial Solution to Optimize Costs, Environmental Requests and Quality for Textile Finishing, Ultrasonics Sonochemistry, 11, 33-38p.

21. http://www.ntcresearch.org/pdf-rpts/anrp98/c95-g13.pdf, 2008

22. Ferrero, F. and Periolatto, M., 2012, Ultrasound for Low Temperature Dyeing of Wool with Acid Dye, Ultrasonics Sonochemistry, 19, 601-606

23. hukla, S.R. and Mathur, M.R., 1995, Low-Temperature Ultrasonic Dyeing of Silk, Journal of the Society of Dyers and Colorists, 111, 342-5p.

24. http://www.fiber2fashion.com, 2008

25. Battu, A., Giansetti, M., Rovero, G. and Sicardi, S., 2010, Intensification of Wet Textile Processing By Ultrasound Application, Proceedings of the 22nd IFATCC International Congress, Stresa, Italy

26. Yukseloğlu, S.M. and Bolat, N., 2010, The Use of Conventional and Ultrasonic Energy in Dyeing of 100% Wool Woven Fabrics, Tekstil ve Konfeksiyon, 2, 162-167p.

27. McNeil, S.J. and McCall, R.A., 2011, Ultrasound for Wool Dyeing and Finishing, Ultrasonics Sonochemistry, 18, 401-406

28. Atav, R., Yurdakul, A., 8-10 June-2011, The Use of Ultrasound in Dyeing of Mohair Fibers, 11th World Textile Conference AUTEX, Moulhouse-France, Book of Abstracts, 277-280p.

29. Srikanth, B., The Basic Benefits of Ultraviolet Technology, http://aquvair.com/Basic%20Benefits%20of%20UV%20-WCP%20Reprint.pdf

30. http://en.wikipedia.org/wiki/Electromagnetic_spectrum, 2012

31. http://skincareclub.wordpress.com/2011/02/25/uva-uvb-uvc-rays/, 2012

32. Bhatti, I.A., Adeel, S. and Abbas, M., Effect of Radiation on Textile Dyeing, http://cdn.intechweb.org/pdfs/25003.pdf

33. http://www.epa.gov/sunwise/doc/uvradiation.html, 2012

34. http://ec.europa.eu/health/opinions2/glossary/tuv/uv-radiation.htm, 2012

35. http://www.google.com.tr/imgres?q=ultraviolet+spectrum&hl=tr&biw =1249&bih=599&tbm=isch&tbnid=lxj5wLLo6i43HM:&imgrefurl=http:// www.uvlp.ca/&docid=OAkRjqeFw_ZM9M&imgurl=http://www.uvlp.ca /images/uvspectrum.jpg&w=450&h=162&ei=3vr7T9XoK8jatAaww8m_ Aw&zoom=1, 2012

36. Treigiene, R. and Musnickas, J. R., 2008, Influence of UV Exposure on Properties of Wool Fiber Pretreated with Surfactants Solutions, Materials Science, 14 (1), 75-78p.

37. Millington, K., 1998, Using Ultraviolet Radiation to Reduce Pilling of Knitted Wool and Cotton, Textile Research Journal, 68 (6), 413-421p.

38. Millington, K. R., CSIRO Textile and Fiber Technology, UV Technology Applications in The Textile Industry, http://www.fiber2fashion. com/industry-article/pdffiles/uv-technology-applications-in-the-textile-industry.pdf?PDFPTOKEN=73534d84ba5e495e4164c5a831509f62275d 95dc%7C1225281847

39. El-Sayed, H. and El-Khatib, E., 2005, Modification of wool fabric using ecologically acceptable UV-assisted treatments, Journal of Chemical Technology and Biotechnology, 80 (10), 1111–1117p.

40. Millington, K. R., 1998, The Use of Ultraviolet Radiation in an Adsorbable Organohalogen-Free Print Preparation for Wool and in Wool Dyeing: the Siroflash Process Journal of the Society of Dyers and Colorists, 114 (10), 286-292p.

41. Millington, K. R., 1998, U.S.A. Patent, Application of UV surface treated in textile industry, Red Tech 98 North Americas: UNEB Conference Proceeding Chicago ILLINOIS America

42. Xin, J.H., Zhu, R., Hua, J. and Shen, J., 2002, Surface Modification and Low Temperature Dyeing Properties of Wool Treated By UV Radiation, Coloration Technology, 118 (4), 169-173p.

43. http://www.sfu.ca/~vwchu/projects/chemistryozone.pdf, 2012

44. Polat, D., 2010, Catalytic Ozonation of Industrial Textile Wastewaters in A Three Phase Fluidized Bed Reactor, Master Thesis, The Graduate School of Natural Sciences of Middle East Technical University

45. http://water.epa.gov/lawsregs/rulesregs/sdwa/mdbp/upload/2001_01_12_mdbp_alter_chapt_3.pdf, 2012

46. Parmenter, K., 2004, Global Ozone Handbook Agriculture and Food Industries http://www.chimia.it/pdf/Ozone%20Handbook%20-%20Revised%2003-11-05%20Final.pdf

47. Balousek, P.J., 1979, The Effects of Ozone Upon A Lignin-Related Model Compound Containing A β-Aryl Ether Linkage, PhD Thesis, Lawrence University

48. http://en.wikipedia.org/wiki/File:Ozone-1,3-dipole.png, 2012

49. http://astrobioloblog.wordpress.com/2011/08/18/life-in-our-solar-system-%E2%80%93-earth/, 2012

50. Iglesias, S.C., 2002, Degradation and Biodegradability Enhancement of Nitrobenzene and 2,4-Dichlorophenol By Means of Advanced Oxidation Processes Based on Ozone, Universitat de Barcelona, PhD thesis

51. Prabaharan, M. and Rao, J.V., 2001, Study on Ozone Bleaching of Cotton Fabric-Process Optimisation, Dyeing and Finishing Properties, Coloration Technolgy, 117 (2), 98-103p.

52. Song, S., Yao, J., He, Z., Qiu, J. and Chen, J., 2008, Effect of Operational Parameters on The Decolorization of C.I. Reactive Blue 19 in Aqueous Solution By Ozone-Enhanced Electrocoagulation, Journal of Hazardous Materials, 152, 204-210p.

53. El-Zaher, N.A. and Micheal, M.N., 2002, Time Optimization of Ultraviolet-Ozone Pretreatment for Improving Wool Fabrics Properties, Journal of Applied Polymer Science, 85(7), 1469-1476p.

54. Atav, R. and Yurdakul, A., 2011, Effect of Ozonation Process on the Dyeability of Mohair Fibers, Coloration Technology, 127(3), 159-166

55. Wu, C.H. and Ng, H.Y., 2008, Degradation of C.I. Reactive Red 2 (RR2) Using Ozone-Based Systems: Comparisons of Decolorization Efficiency and Power Consumption, Journal of Hazardous Materials, 152, 120-127p.

56. Eren, H.A. and Aniş, P., 2006, Decolorisation of Textile Effluents by Ozonation, Uludağ University Journal of the Faculty of Engineering and Architecture, 11(1), 83-91p.

57. Perincek, S., Bahtiyari, M.I., Körlü, A.E. and Duran, K., 2008, Ozone Treatment of Angora Rabbit Fiber, Journal of Cleaner Production, 16, 1900-1906p.

58. Sargunamani, D. and Selvakumar, N., 2007, Effects of Ozone Treatment on The Properties of Raw And Degummed Tassar Silk Fabrics, Journal of Applied Polymer Science, 104 (1), 147–155p.

59. Dereli Perincek, S., 2006, An investigation on The Applicability of Ultrasound, Ultraviolet, Ozone and Combination of These Technologies as A Pretreatment Process, Master Thesis, Ege Üniversitesi Fen Bilimleri Enstitüsü

60. Micheal, M.N. and El-Zaher, N.A., 2003, Efficiency of Ultraviolet/Ozone Treatments in The Improvement of The Dyeability and Light Fastness Of Wool, Journal of Applied Polymer Science, 90 (13), 3668-3675p.

61. Atav, R. and Yurdakul, A., 2010, Low Temperature Dyeing of Ozonated Angora Fibers, 10th World Textile Conference AUTEX, Vilnius-Litvania, Book of Abstracts, 16p.

62. Sparavigna, A., Plasma Treatment Advantages for Textiles, http://arxiv.org/ftp/arxiv/papers/0801/0801.3727.pdf

63. Moore, R., Plasma Surface Functionalization of Textiles, http://www.acteco.org/Acteco/public/results/MOORE_IoN_March08.pdf

64. http://www.britannica.com/EBchecked/media/148660/States-of-matter, 2012

65. http://www.fiber2fashion.com/industry-article/17/1636/plasma-treatment-technology-for-textile-industry1.asp, 2012

66. http://www-alt.igb.fraunhofer.de/www/gf/grenzflmem/schichten/ en/ TechTextile.en.html, 2012

67. Shyam Sundar, P., Prabhu, K.H. and Karthikeyan, N.,2007, Fourth State Treatment for Textiles, The Indian Textile Journal, http://www. indiantextilejournal.com/articles/FAdetails.asp?id=290

68. Desai, A.A., 2008, Plasma Technology: A Review, The Indian Textile Journal http://www.indiantextilejournal.com/articles/ FAdetails. asp?id=775

69. Buyle, G., 2009, Nanoscale Finishing of Textiles Via Plasma Treatment, Materials Technology, 24(1), 46-51p., http://www.centexbel.be/files/ publication-pdf/MTE_Buyle.pdf

70. Karahan, H.A., Özdoğan, E., Demir, A., Ayhan, H. and Seventekin, N., 2009, Effects of Atmospheric Pressure Plasma Treatments on Certain Properties of Cotton Fabrics, Fibers & Textiles in Eastern Europe, 17 (2), 19-22p.

71. Karahan, A., Yaman, N., Demir, A., Ozdoğan, E., Oktem, T. and Seventekin, S., 2006, Tekstilde Plazma Teknolojisinin Kullanım Olanakları: Plazma Nedir?, Tekstil ve Konfeksiyon, 16(1),302-309p.

72. Jocic, D., Vilchez, S., Topalovic, T., Molina, R., Navarro, A., Jovancic, P., Julia, M.R. and Erra, P., 2005, Effect of Low-Temperature Plasma and Chitosan Treatment on Wool Dyeing with Acid Red 27, Journal of Applied Polymer Science, 97(6), 2204-2214p.

73. http://www.tno.nl/downloads/def_maritiem_plasmaapplicaties_DV2_ 05d046.pdf, 2012

74. Ahmed, A., 2007, A Review of Plasma Treatment for Application on Textile Substrate, BSc. Thesis, The Faculty of Textile Science

75. Erra, P., Molina, R., Jocic, D., Julia, M.R., Cuesta, A., and Tascon, J.M.D., 1999, Shrinkage Properties of Wool Treated with Low Temperature Plasma and Chitosan Biopolymer, Textile Research Journal, 69(11), 811-815p.

76. Höcker, H., 2002, Plasma Treatment of Textile Fibers, Pure Appl. Chem., 74(3), 423-427p.

77. Wakida, T., Tokino, S., Niu, S., Lee, M., Uchiyama, H. and Kaneko, M., 1993, Dyeing Properties of Wool Treated with Low-Temperature Plasma Under Atmospheric Pressure, Textile Research Journal, 63(8), 438-442p.

78. Yoon, N.M., Lim, Y.J., Tahara, M. and Takagishi, T., 1996, Mechanical and Dyeing Properties of Wool and Cotton Fabrics Treated with Low Temperature Plasma and Enzymes, Textile Research Journal, 66(5), 329-336p.

79. Jing, Z., 1996, Silk Fabrics Modification Through Plasma Graft Copolymerization, Journal of Textile Research, 4

80. Wakida, T., Lee, M., Sato, Y., Ogasawara, S., Ge, Y. and Niu, S., 1996, Dyeing Properties of Oxygen Low-Temperature Plasma-Treated Wool and Nylon 6 Fibers with Acid And Basic Dyes, Journal of the Society of Dyers and Colorists, 112(9), 233-236p.

81. Wakida, T., Cho, S., Choi, S., Tokino, S. and Lee, M., 1998, Effect of Low Temperature Plasma Treatment on Color of Wool and Nylon 6 Fabrics Dyed with Natural Dyes, Textile Research Journal, 68(11), 848-853p.

82. Kan, C.W., Chan, K., Yuen, C.W.M. and Miao, M.H., 1998, Surface Properties of Low-Temperature Plasma Treated Wool Fabrics, Journal of Materials Processing Technology, 83(1-3), 180-184p.

83. Kan, C.W., Chan, K., Yuen, C.W.M. and Miao, M.H., 1998, The Effect of Low-Temperature Plasma on The Chrome Dyeing of Wool Fiber, Journal of Materials Processing Technology, 82(1-3), 122-126p.

84. Kan, C.W., Chan, K., Yuen, C.W.M. and Miao, M.H., 1999, Low Temperature Plasma on Wool Substrates: The Effect of The Nature of The Gas, Textile Research Journal, 69(6), 407-416p.

85. Kan, C.W., Chan, K. and Yuen, C.W.M., 2004, Surface Characterization of Low Temperature Plasma Treated Wool Fiber-The Effect of The Nature of Gas-, Fibers and Polymers, 5(1), 52-58p.

86. Iriyama, Y., Mochizuki, T., Watanabe, M. and Utada, M., 2002, Plasma Treatment of Silk Fabrics for Better Dyeability, Journal of Photopolymer Science and Technology, 15(2), 299-306p.

87. Sun, D. and Stylios, G.K., 2004, Effect of Low Temperature Plasma Treatment on The Scouring and Dyeing of Natural Fabrics, Textile Research Journal, 74(99), 751-756p.

88. Biniaś, D., Włochowicz, A. and Binias, W., 2004, Selected Properties of Wool Treated By Low-Temperature Plasma, Fibers&Textiles in Eastern Europe, 12(2), 58-62p.

89. Sun, D. and Stylios, G.K., 2005, Investigating The Plasma Modification of Natural Fiber Fabrics-The Effect on Fabric Surface and Mechanical Properties, Textile Research Journal, 75(9), 639-644p.

90. Masukuni, M. and Norihiro, I., 2006, Dyeing Properties of Argon-Plasma Treated Wool, Sen'I Gakkaishi, 62(9), 205-211p.

91. Kan, C. and Yuen, C.M., 2006, Dyeing Behavior of Low Temperature Plasma Treated Wool, Plasma Processes and Polymers, 3 (8), 627-635p.

92. El-Zawahry, M.M., Ibrahim, N.A. and Eid, M.A., 2006, The Impact of Nitrogen Plasma Treatment upon the Physical-Chemical and Dyeing Properties of Wool Fabric, Polymer-Plastics Technology and Engineering, 45(10), 1123-1132p.

93. Demir, A., Karahan, A., Ozdogan, E., Oktem, T. and Seventekin, N., 2008, The Synergetic Effects of Alternative Methods in Wool Finishing, Fibers&Textiles in Eastern Europe, 16(2), 89-94p.

94. Chvalinova, R. and Wiener, 2008, Sorption Properties of Wool Fibers After Plasma Treatment, II Central European Symposium on Plasma Chemistry, Chem. Listy 102, 1473-1477p.

95. Naebe, M., Cookson, P. G., Rippon, J., Brady, R. P., Wang, X., Brack, N. and van Riessen, G., 2010, Effects of Plasma Treatment of Wool on the Uptake of Sulfonated Dyes with Different Hydrophobic Properties, Textile Research Journal, 80(4), 312-324p.

96. Demir A., 2010, Atmospheric Plasma Advantages for Mohair Fibers in Textile Applications, Fibers and Polymers, 11(4), 580-585p.

97. Atav, R. and Yurdakul, A., 2011, Low Temperature Dyeing of Plasma Treated Luxury Fibers. Part I: Results for Mohair (Angora Goat), Fibers & Textiles in Eastern Europe, 19(2), 84-89p.

98. http://en.wikipedia.org/wiki/Gamma_ray, 2012

99. http://web.princeton.edu/sites/ehs/osradtraining/radiationproperties/radiationproperties.htm, 2012

100. Zohdy, M.H., El-Naggar, A.M. and Marie, M.M., 1999, Effect of Copper Treatment on The Dyeability of Gamma Irradiated Acrylic Fabrics with Different Dyestuffs, Materials Chemistry and Physics, 61, 237-243p.

101. Aytac, A., Sen, M., Deniz, V. and Guven, O., 2007, Effect of Gamma Irradiation on The Properties of Tyre Cords, Nuclear Instruments and Methods in Physics Research, 265, 271-275p.

102. Zohdy, M.H., El-Naggar, A.M. and Abdallah, W.A., 1997, Silk Screen Printing of Some Reactive Dyes on Gamma Irradiated Wool Fabrics, Polymer Degradation and Stability, 55, 185-189p.

103. Horio, M., Ogami, K., Kondo, T. and Sekimoto, K., 1962, Effect of Gamma Irradiation upon Wool Fibers, http://repository.kulib.kyoto-u.ac.jp/dspace/bitstream/2433/75949/1/chd041_1_001.pdf

104. Beevers, R.B. and McLaren, K.G., 1974, The Effect of Low Doses of Co60 Gamma Radiation on Some Physical Properties and The Structure of Wool Fibers, Textile Research Journal, 44(12), 986-994p.

105. Millington, K.R., 2000, Comparison of The Effects of Gamma and Ultraviolet Radiation on Wool Keratin, Journal of The Society of Dyers and Colourists, 116, 266-272p.

106. Kan, C.W., 2008, Impact on Textile Properties of Polyester with Laser, Optics & Laser Technology, 40, 113-119p.

107. http://en.wikipedia.org/wiki/Laser, 2012

108. ttp://resources.alibaba.com/topic/800007870/Laser_processing_technology_in_Textile_and_Apparel.htm, 2012

109. Montazer, M., Taheri, S.J. and Harifi, T., 2011, Effect of Laser CO2 Irradiation on Various Properties of Polyester Fabric: Focus on Dyeing, Journal of Applied Polymer Science, 124(1), 342-348p.

110. Esteves, F. and Alonso, H., 2007, Effect of CO2 Laser Radiation on Surface and Dyeing Properties of Synthetic Fibers, Research Journal of Textile and Apparel, 11(3), 42-47p.

111. http://en.wikipedia.org/wiki/Excimer_laser, 2012

112. Knittel, D. and Schollmeyer, E., 1998, Surface Structuring of Synthetic Fibers by UV Laser Irradiation. Part III. Surface Functionality Changes Resulting from Excimer-Laser Irradiation, Polymer International, 45, 103-109p.

113. Nourbakhsh, S. and Ebrahimi, I., 2012, Different Surface Modification of Poly (Ethylene Terephthalate) and Polyamide 66 Fibers by Atmospheric Air Plasma Discharge and Laser Treatment: Surface Morphology and Soil Release Behavior, Journal of Textile Science&Engineering, 2(2), http://dx.doi.org/10.4172/2165-8064.1000109

114. Bahtiyari, M.I., 2011, Laser Modification of Polyamide Fabrics, Optics & Laser Technology, 43(1), 114-118p.

115. Yip, J., Chan, K., Sin, K. and Lau, K., 2002, UV Excimer Laser Modification on Polyamide Materials: Effect on The Dyeing Properties, Materials Research Innovations, 6(2), 73-78p.

116. Buyukakıncı, B.Y., 2012, Usage of Microwave Energy in Turkish Textile Production Sector, Energy Procedia, 14, 424-431p.

117. http://en.wikipedia.org/wiki/Microwave, 2012

118. http://www.gallawa.com/microtech/mwave.html, 2012

119. http://www.microwavetec.com/theor_basics.php, 2012

120. Ahmed, N.S.E and El-Shishtawy, R.M., 2010, The Use of New Technologies in Coloration of Textile Fibers, Journal of Material Science, 45(5), 1143-1153p.

121. http://www.rsc.org/images/evaluserve_tcm18-16758.pdf, 2012

122. Collins, J.M. and Leadbeater, N.E., 2007, Microwave Energy: A Versatile Tool for The Biosciences, Organic & Bimolecular Chemistry, 5, 1141-1150p.

123. Haggag, K., Hanna, H.L., Youssef, B.M. and El-Shimy, N.S., March 1995, Dyeing Polyester With Microwave Heating Using Disperse Dyestuffs, American Dyestuff Reporter, 22-35p.

124. Delaney, M.J. and Seltzer, I., 1972, Microwave Heating for Fixation of Pad-dyeings on Wool, Journal of the Society of Dyers and Colorists, 88(2), 55-59p.

125. Zhao, X. and He, J.X., 2011, Improvement in Dyeability of Wool Fabric By Microwave Treatment, Indian Journal of Fiber & Textile Research, 36(1), 58-62p.

126. Leeder, J.D. , McGregor, B.A. and Steadman, R.G., 1998, Properties and Performance of Goat Fiber, RIRDC Publication, No 98/22, RIRDC, Proje No: ULA-8A

127. Machi, S., 1996, New Trends of Radiation Processing Applications, Radiation Physics and Chemistry, 47(3), 333-336p.

128. Li, J., Huang, Y., Xu, Z. and Wang, Z., 2005, High-Energy Radiation Technique Treat on The Surface Of Carbon Fiber, Materials Chemistry and Physics, 94, 315-321p.

129. Alberti, A., Bertini, S., Gastaldi, G., Iannaccone, N., Macciantelli, D., Torri, G. and Vismara, E., 2005, Electron Beam Irradiated Textile Cellulose Fibers. ESR Studies and Derivatisation with Glycidyl Methacrylate (GMA), European Polymer Journal, 41, 1787-1797p.

130. Fatarella, E., Ciabatti, I. and Cortez, J., 2010, Plasma and Electron-Beam Processes as Pretreatments for Enzymatic Processes, Enzyme and Microbial Technology, 46, 100-106p.

131. Vahdat, A., Bahrami, H., Ansari, N. and Ziaie, F., 2007, Radiation Grafting of Styrene onto Polypropylene Fibers by A 10mev Electron Beam, Radiation Physics and Chemistry, 76, 787-793p.

132. Clough, R.L., 2001, High-Energy Radiation and Polymers: A Review of Commercial Process and Emerging Applications, Nuclear Instruments and Methods in Physics Research B, 185, 8-33p.

133. Chmielewski, A.G. and Haji-Saeid, M., 2004, Radiation Technologies: Past, Present and Future, Radiation Physics and Chemistry, 71, 16-20p.

134. Oktem, T., Ozdogan, E., Namligoz, S.E., Oztarhan, A., Tek, Z., Tarakcıoglu, I. and Karaaslan, A., 2006, Investigating the Applicability of

Metal Ion Implantation Technique (MEVVA) to Textile Surfaces, Textile Research Journal, 76(1), 32-40p.

135. Ion Beam Processing (IBP) Technologies, Sector Study, FINAL REPORT, Prepared for the North American Technology and Industrial Base Organization (NATIBO), Prepared by BDM Federal, June 1996

136. Dong, H. and Bell, T., 1999, State-of-The-Art Overview: Ion Beam Surface Modification of Polymers Towards Improving Tribological Properties, Surface and Coatings Technology, 111, 29-40p.

137. Montero, G., Hinks, D. and Hooker, J., 2003, Reducing Problems of Cyclic Trimer Deposits in Supercritical Carbon Dioxide Polyester Dyeing Machinery, Journal of Supercritical Fluids, 26, 47-54p.

138. Fernandez Cid, M.V., van Spronsen, J., van der Kraan, M., Veugelers, W.J.T., Woerlee, G.F. and Witkamp, G.J., 2007, A significant approach to dye cotton in supercritical carbon dioxide with fluorotriazine reactive dyes, J. of Supercritical Fluids, 40, 477-484p.

139. Hou, A. and Dai, J., 2005, Kinetics of Dyeing of polyester with CI Disperse Blue 79 in Supercritical Carbondioxide, Coloration Technology, 121, 18-20p.

140. Özcan, A.S., Clifford, A.A., Bartle, K.D. and Lewis, D.M., 1998, Dyeing of Cotton Fibres with Disperse Dyes in Supercritical Carbon Dioxide, Dyes and Pigments, 36 (2), 103-110p.

141. New Technologies in Textile Dyeing and Finishing, http://www.industryhk.org/english/fp/fp_hki/files/HKI_03_09_textile_e.pdf

142. Özcan, A.S., and Özcan, A., 2005, Adsorption Behavior of A Disperse Dye on Polyester in Supercritical Carbon Dioxide, Journal of Supercritical Fluids, 35, 133-139p.

143. Hou, A., Chen, B., Dai, J. and Zhang, K., 2010, Using Supercritical Carbon Dioxide as Solvent to Replace Water in Polyethylene Terephthalate (PET) Fabric Dyeing Procedures, Journal of Cleaner Production, 18,1009-1014p.

144. Jun, J.H., Ueda, M., Sawada, K., Sugimotod, M. and Urakawaa, H., 2005, Supercritical Carbon Dioxide Containing A Cationic

Perfluoropolyether Surfactant for Dyeing Wool, Coloration Technology, 121, 315-319p.

145. Özcan, A.S., Clifford, A.A., Bartle, K.D., Broadbent, P.J. and Lewis, D.M., 1998, Dyeing of Modified Cotton Fibres with Disperse Dyes from Supercritical Carbon Dioxide, Journal of The Society of Dyers and Colourists, 114, 169-173p.

146. Van der Kraan, M., Bayrak, Ö., Fernandez Cid, M.V., Woerlee, G.F., Veugelers, W.J.T. and Witkamp, G.J., Textile Dyeing in Supercritical Carbon Dioxide, http://www.isasf.net/fileadmin/files/Docs/Versailles/Papers/PMt2.pdf

147. Guzel, B. and Akgerman, A., 2000, Mordant Dyeing of Wool By Supercritical Processing, Journal of Supercritical Fluids, 18, 247-252p.

148. Sawada, K., Takagi, T., Jun, J.H., Ueda, M. and Lewis, D.M., 2002, Dyeing Natural Fibres in Supercritical Carbon Dioxide Using A Nonionic Surfactant Reverse Micellar System, Coloration Technology, 118(5), 233-237p.

149. Jun, J.H., Sawada K. and Ueda, M., 2004, Application of Perfluoropolyether Reverse Micelles in Supercritical CO2 to Dyeing Process, Dyes and Pigments, 61, 17-22p.

150. Van der Kraan, M., Fernandez Cid, M.V., Woerlee, G.F., Veugelers, W.J.T. and Witkamp, G.J., 2007, Dyeing of Natural and Synthetic Textiles in Supercritical Carbon Dioxide with Disperse Reactive Dyes, The Journal of Supercritical Fluids, 40, 470-476p.

Chapter 5

Dyeing and thermal behavior of jute fibre grafted with nitrile monomer

Md Ibrahim H Mondal , Md Khademul Islam

ABSTRACT

Modification of bleached jute fibre was carried out by graft co-polymerization with the vinyl monomers, acrylonitrile and methacrylonitrile. A $K_2S_2O_8/FeSO_4$ redox system was used in the presence of air. The graft yield and grafting efficiency increased with the increase of monomer, initiator and reaction time up to certain values, thereafter it decreased. This is due to the increasing rate of homopolymerization, rather than grafting, at the higher monomer concentration. The maximum graft yield with acrylonitrile and methacrylonitrile, under optimized conditions, was 19.78 and 43.15%, respectively. The graft yield of methacrylonitrile is greater than that of acrylonitrile. This is due to the $-CH_3$ group present in methacrylonitrile. The inclusion of nitrile monomer onto the treated fibre was identified by Fourier Transform Infrared Spectroscopy. The surface morphology and thermal behavior of the modified fibre was investigated by Scanning Electron Microscopy and Thermogravimetric Analysis. Modification of jute fibre improved the thermal stability, breaking strength as well as dyeability of the fiber.

KEYWORDS

Jute, Grafting, Nitrile monomer, Thermal stability

1. INTRODUCTION

Jute fibre is the most important cash crop and the biggest foreign exchange earner of Bangladesh. It is composed of 60–70% α-cellulose, 15–20% hemicelluloses, 5–10% lignin and 3–5% pectins, together with some fatty acid and waxy matter (Alam and Khan 2007). In recent years, it has been threatened by increasing competition from cheap synthetic fibres. Due to environmental issues, many ecofriendly alternative fibres and chemical modifications have been considered by researchers (Mishra et al. 2001; Gonaslves and Mungara 1996). So it is necessary to improve the quality of jute fibre for commercial utilization, to meet total fabric requirements of the importing countries as well as to minimize the import of competing synthetic fibres.

Various chemical treatments have been used to improve the physico-chemical performance of natural fibres in the past (Ouajai et al. 2007; Mohanty et al. 2000; Dong et al. 2015). Although a lot of work has been reported on graft co-polymerization of vinyl monomers onto cellulosic and other textile fibre (Mondal2013; Bakr et al. 2015), much less has been reported on grafting of nitrile monomers onto jute fibre. Yet jute is an important lignocellulose fibre.

Depending on the chemical structure of the monomer grafted onto cellulose, graft copolymer gain new properties such as improved elasticity, hydrophobic character (Bicak et al. 1999; Okieimen 1987), heat resistance (Samal et al. 1988; Misra et al.1987; Huque et al. 1980), resistance to microbial attack (Mcdowall et al. 1984), abrasion resistance (Bianchi et al. 2000). Grafting also changes the mechanical properties of grafted polymer varied with the amount of grafting (Ouajai et al. 2004; Fatimah et al. 2014; El-Shekeil et al. 2014) and the grafting led little degradation effect on mechanical properties. Hence, it is expected that the properties of this fibre on modification by grafting, can be improved for use in the manufacture of high-quality products.

Therefore chemical modification, through graft co-polymerization, of nitrile monomer on to jute fibres, has been receiving considerable interest in recent years. The grafting was determined on the basis of the increased weight of the fibre treated. In the present work, attempts have been made to modify the jute fibre with acrylonitrile (AN) and methacrylonitrile (MAN) monomers, initiated under a potassium persulphate ($K_2S_2O_8$) and ferrous sulphate ($FeSO_4$) redox system. The study attempted to determine the optimum grafting conditions for such modification. Some physico-chemical characteristics of the grafted fibre were investigated.

2. METHODS

2.1 Materials

Raw jute fibre (*Corchorus olitorius, Tossa Variety*) was collected from Rajshahi Jute Mill Ltd., Bangladesh. Glacial acetic acid, anhydrous sodium acetate, sodium chlorite, acrylonitrile, methacrylonitrile, potassium persulphate, ferrous sulphate and the dyestuffs were purchased from BDH, England.

2.2 Sample preparation

The collected jute fibre were carefully combed, blended and washed with 6.5 g of detergent and 3.5 g of soda per litre at 75°C for 30 min (Farouqui and Mondal1989). The jute fibre was bleached, with a sodium chlorite solution of concentration 0.5%, at pH 4 and at 80–90°C, for 90 min (Farouqui and Mondal 1989). Then the bleached jute fibre was grafted with acrylonitrile and methacrylonitrile monomers.

2.3 Method of grafting

The graft copolymerization of jute fibre was carried out in a 100 mL stoppered Erlenmeyer flask. Polymerization was done with 240–330% monomer, 18–25% $K_2S_2O_8$ as initiator and 1–6% $FeSO_4$ as catalyst, based on the weight of the fibre. Polymerization took 30–150 min, at 60–110°C in the fibre–liquor ratio of 1:50. At the end of the desired period, the jute fibre was washed repeatedly three times with warm distilled water and dried

(Hebiesh et al. 1971; EL-Rafie et al. 1974). Graft yield and grafting efficiency were calculated according to the following formula:

$$\text{Graft yield, \%} = \frac{W_1 - W_0}{W_0} \times 100$$

$$\text{Grafting efficiency, \%} = \frac{W_1 - W_0}{W_2} \times 100$$

where W_1 is the weight of the grafted jute fibre after modification, W_0 is the weight of ungrafted jute fibre before modification and W_2 is the weight of monomer used.

2.4 IR spectroscopy

Fourier Transform Infrared Spectroscopy (FTIR) spectroscopy analysis was performed, using a Perkin Elmer 100 infrared spectrometer. The fibres and KBr (potassium bromide) were dried in an oven at 105°C, until they were moisture-free. Then about one percent fibre was mixed with dried KBr, ground into powder, using a mortar pestle (Mondal 2013). The infrared spectra of the samples were recorded with FTIR between 400 and 4,000 cm^{-1}.

2.5 Thermal analysis

Thermogravimetric analysis was used to characterize the thermal decomposition rate and the thermal stability of the raw and grafted fibre. The samples, approximately 10 mg each, were heated from 30 to 600°C. Heating was done in an inert atmosphere (argon), at a rate of 20°C/min, in a Seiko-extar-TG/DTA-6300, Tokyo, Japan.

2.6 Scanning electron microscopy

Scanning electron microscopy (SEM) was performed. An SEM (FEI Quanta Inspect, Model: S50, Kyoto, Japan was used to observe the surface microstructure of the treated and untreated fibres.

2.7 Method of dyeing

Dyeing of bleached and modified jute fibres was carried out with direct and reactive dyes. The dye concentration of reactive and direct dyes were 3 and 1%, and the electrolyte concentration for the dyes were 7 and 25% respectively.

2.7.1 Tensile strength measurement

The tensile strength of the jute fibres was measured, using a tensile strength tester (Serigraph machine 4004, Japan). Jute fibres were cut into equal pieces, 30 cm in length, and the length of each specimen between the jaws was maintained at 10 cm. One twist per 2 cm was given, along the length of the fibre between the jaws of the machine. The average breaking load (the mean of 10 readings) of the fibre was measured in kg/Yarn. Tenacity was calculated by the formula 9 (ISO 5081-1977(E) 1997).

$$\text{Tenacity} = \frac{\text{Average breaking load}}{\text{Denier}} \times \frac{\text{g}}{\text{Denier}}$$

where,

$$\text{Denier} = \frac{\text{Weight of the sample in grams}}{\text{Length of the sample in meters}} \times 9{,}000.$$

2.7.2 Method of sunlight exposure

The jute sample was exposed directly, on a flat board, to sunlight, without any protection from weathering but was protected from rain, snow etc. At the same time and in the same place, bleached and grafted jute fibres were exposed to sun on the roof of a building in the months of April to July for 7 h each day. This continued for a total of 350 h.

2.7.3 Method of heating

Jute sample was placed in an electric oven, in the presence of air, at 30–160°C for 3 h. Its breaking strength was then measured as described above.

3. RESULTS AND DISCUSSION

3.1 Optimization of graft copolymerization

The effect of the monomer concentration (AN and MAN) on polymerization of bleached jute fibre in the presence of a $K_2S_2O_8/FeSO_4$ redox system, is shown in Fig. 1. The graft yield increased with the increase of monomer concentration for both AN and MAN. After a certain value, it decreased with a further increase in the monomer concentration (Samal et al. 1986, 1989). The latter is due to the increasing homopolymerization rate rather than to grafting at the higher monomer concentration. The maximum graft yield was obtained when the bleached jute fibre was treated with 300% AN and 280% MAN. Their grafting efficiencies are 4.50 and 11.00%, respectively.

Figure 2 shows that the graft yield and grafting efficiency increased with the increase of $K_2S_2O_8$ concentration as initiator, up to 22% for AN and 20% for MAN. The increasing trend of the graft yield might be the result of the formation of a large number of grafting sites on the cellulose backbone of the jute fibre. At the high initiator concentration, a large number of activated monomers become available. These turn to the formation of homopolymers, causing a shortage of monomer in the vicinity of the cellulose macroradicals. Hence, the graft yield and grafting efficiency start to decrease. The retarding effects of graft yield and grafting efficiency at higher concentration of $K_2S_2O_8$ may be due to the predominance of homopolymerization over grafting. The termination of growing grafted chain is caused primarily by free radicals, resulting from the decomposition of excess initiator and production of excess Fe(III) (Mukhopadhyay et al. 1975; Misra et al.1987; Mohanty et al. 1986).

Figure 3 shows that the percent graft yield increased with the increase of $FeSO_4$concentration, up to 2%, for both AN and MAN. Thereafter, percent graft yield decreased. As the catalyst concentration increases, a greater number of Fe(II) ions is produced. Also, more reactive sites on the jute surface develop. Hence, grafting increases up to a certain concentration of catalyst. Beyond this concentration, graft yield and grafting efficiency decline. This is probably due to the formation of Fe(III) ions, which promote the premature termination of the growing grafted chain. Homopolymerization is also enhanced in presence of $FeSO_4$ (Salam 2005). The effective concentrations of $FeSO_4$ are 2% for both AN and MAN, and the corresponding graft yield and grafting efficiency rates are 17. 78, 43.15 and 6.5, 14.38% respectively.

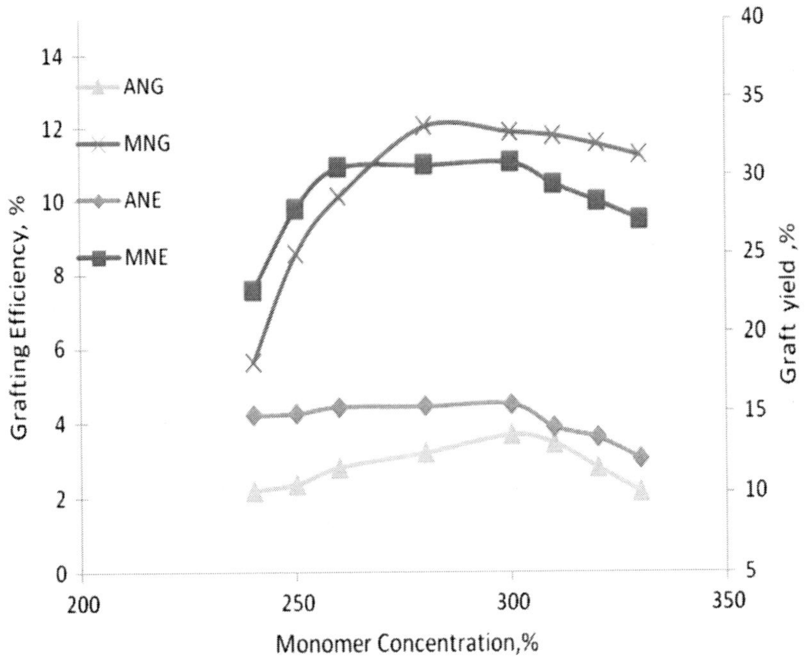

Figure1. Graft yield and grafting efficiency as a function of monomer concentration (*ANG* acrylonitrile grafting, *MNG*methacrylonitrile grafting, *ANE* efficiency of acrylonitrile, *MNE* efficiency of methacry-lonitrile).

Figure 4 shows that the graft yield increased with the increase of reaction time, up to 90 min for both AN and MAN. During this span of time, the activated monomer combined with the active sites of the fibre matrix. After 90 min, the monomers found more reactive sites on the fibre backbone of cellulose and thus became homopolymerized (EL-Rafie 1994).

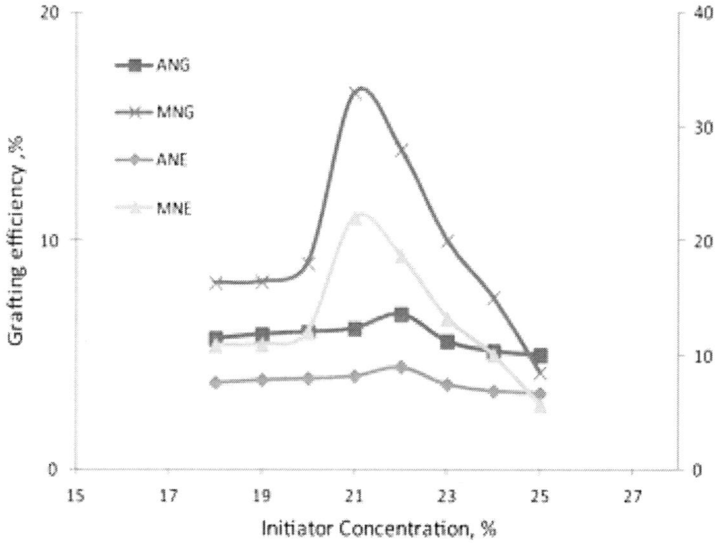

Figure 2. Graft yield and grafting efficiency as a function of initiator concentration (*ANG* acrylonitrile grafting, *MNG*methacrylonitrile grafting, *ANE* efficiency of acrylonitrile, *MNE* efficiency of methacryl-onitrile).

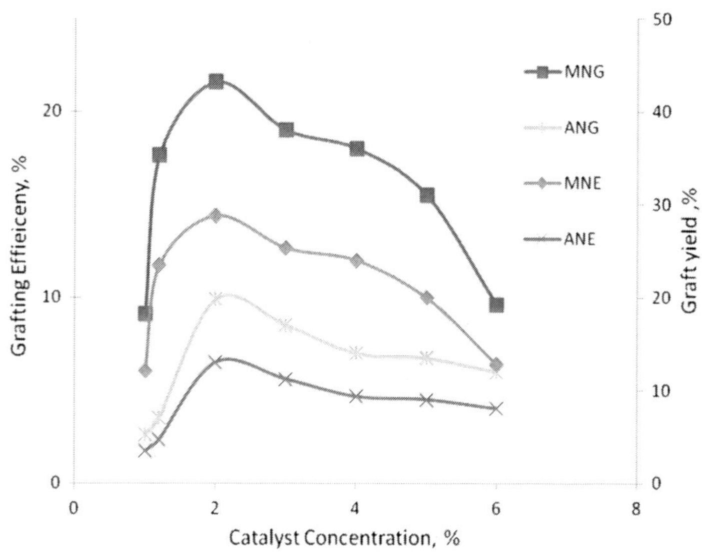

Figure 3. Graft yield and grafting efficiency as a function of catalyst ($FeSO_4$) concentration (*ANG* acrylonitrile grafting,*MNG* methacrylonitrile grafting, *ANE* efficiency of acrylonitrile, *MNE* efficiency of methacry-lonitrile).

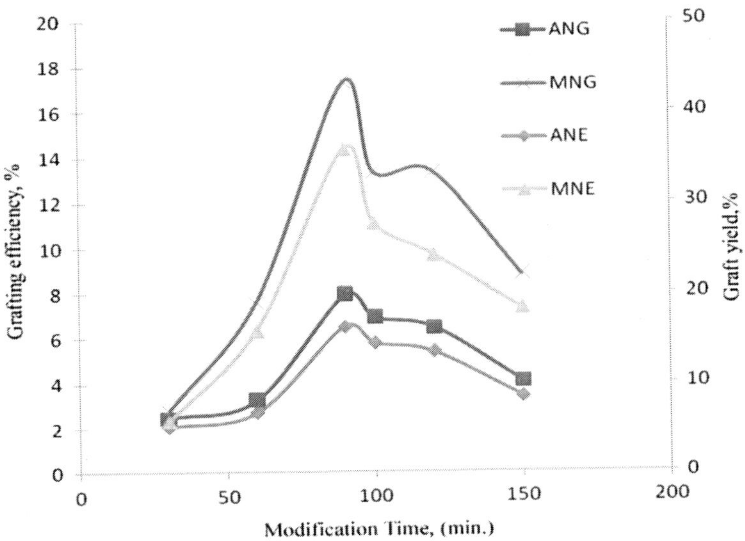

Figure 4. Graft yield and grafting efficiency as a function of reaction time (*ANG* acrylonitrile grafting, *MNG*methacrylonitrile grafting, *ANE* efficiency of acrylonitrile, *MNE* efficiency of methacrylonitrile).

Again, at longer reaction times, the graft yield decreased. This could be due to the partial dissolution of the grafted fibre (Hebeish and Mehata 1968). The optimum polymerization time was 90 min for both AN and MAN, while the corresponding graft yields were 19.78 and 43.15% respectively.

The effect of polymerization temperature on grafting of AN and MAN onto bleached jute fibre is shown in Fig. 5. It is seen that the grafting rate is low at low temperature, but at 80°C, the graft yield and grafting efficiency peaked. The graft yield and grafting efficiency are 19.78, 6.59% for AN and 43.15, 14.38% for MAN, respectively, within an 80–90°C temperature range. The increase in graft yield and grafting efficiency up to optimum temperature may be ascribed as increase in the rate of production of active free radicals. These free radicals increase the number of grafting sites on the fibre backbone at a higher rate. Increase of temperature increases the rate of diffusion of monomer into the jute matrix, where grafting is also initiated by complex monomers (Rahman 1970).

The decrease in grafting efficiency and graft yield, beyond optimum temperature, may be attributed to the increase in activation energy for graft copolymerization. As discussed above, premature termination of growing grafted chains by excess Fe(III) ions, produced on the oxidation of Fe(II) ions, can also be a factor,

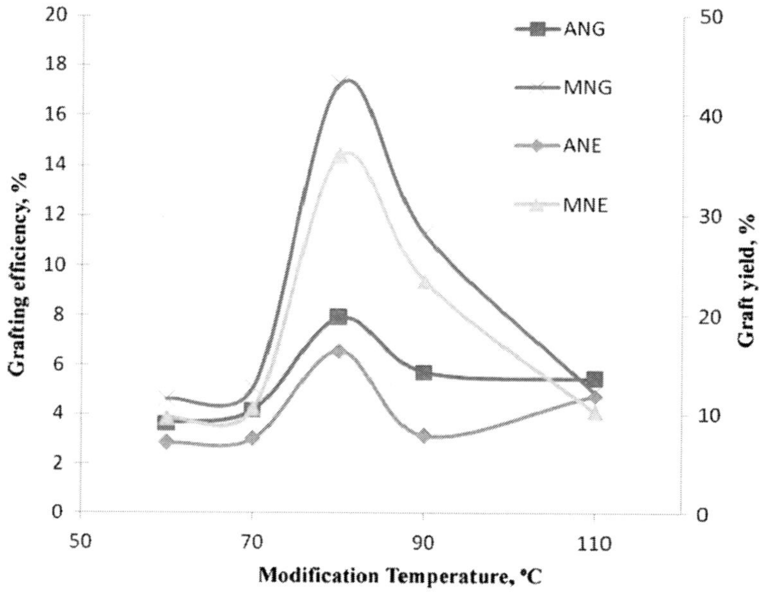

Figure 5. Graft yield and grafting efficiency as a function of reaction temperature (*ANG* acrylonitrile grafting, *MNG*methacrylonitrile grafting, *ANE* efficiency of acrylonitrile, *MNE* efficiency of methacrylonitrile).

3.2 Dyeing behavior of the grafted jute fibre

The exhaustion of direct dyes by bleached and grafted jute fibres is listed in Table 1. The dye exhaustion of the bleached jute fibre was higher than that of the modified fibre. The dye exhaustion decreased with an increase in the percent graft yield which occurred due to increase in hydrophobicity of grafted fibre. The dye absorption of fibre depends on the availability of the total external surface of the fibre pores or cavities in a fixed amount of the fibre and the attractive force between the fibres and dye ion. The bleached jute fibre had more available pores or cavities than the modified jute fibre as these, on grafting, were blocked. Hence, the dye exhaustion of the grafted jute fibre was less than that of the bleached jute fibre. The hydrophobic nature of the grafted jute fibre may also have increased with increases in the grafting amount. In such a case, the dye exhaustion of the bleached jute fibre would be higher than that of the grafted jute fibre.

Table 1. Effect of dye absorption on dyeing of bleached and modified (AN and MAN) jute fibres

Dyes	Name of dye	Dye exhaustion, %		
		Bleached fibre	AN-grafted fibre	MN-grafted fibre
Reactive dyes	Reactive Orange 14	47.18	46.00	44.4
	Reactive Brown 10	46.40	40.28	38.23
Direct dyes	Direct Orange 31	83.18	82.40	81.13
	Direct Blue 1	94.15	90.38	89.23

3.3 FTIR spectra of grafted jute fibre

The infrared spectra of the bleached and grafted jute fibres are shown in the Fig. 6. The IR spectra of bleached and grafted fibre were more or less similar. The FTIR spectra of the bleached and grafted jute fibre exert absorption peaks at 3,200–3,600 cm^{-1} due to the –OH vibration, 2,820–3,050 cm^{-1} result from the C–H stretching vibration, and the peak at 1,163 cm^{-1} due to the antisymmetric bridge C–O–C stretching's vibration of the cellulose pyranose ring. The presence of an absorption band near 1,711 cm^{-1} in the bleached jute fibre was probably due to the C=O stretching of the ester group in celluloses (Tsuboi 1957). The new additional peaks in the grafted jute fibre were appeared at around 2,240 and 2,236 cm^{-1} for AN and MAN respectively indicated the presence of nitrile (–C≡N) group (Mondal et al.2004; Khullar et al. 2008; Patra and Singh 1994).

3.4 Determination of moisture content

The moisture content of bleached fibre is 7.3% and that of modified fibre with AN and MAN fibres are 2.84 and 4.01%, at 105°C, respectively. These were measured by a moisture analyzer (Model: R.A 120-3, Kern, Germany).

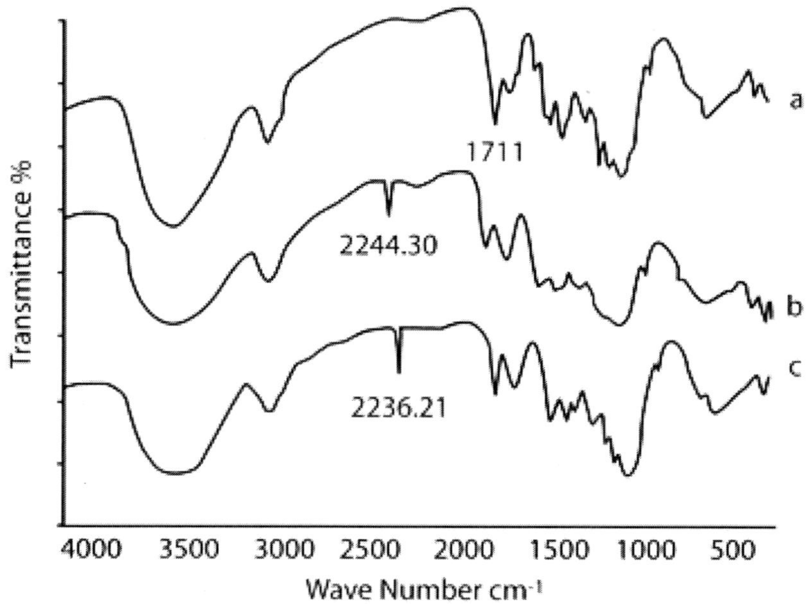

Figure 6. FTIR spectra of (*a*) bleached jute, (*b*) acrylonitrile and (*c*) methacrylonitrile modified jute fibre.

3.5 Thermal stability

Figure 7a–c represents the thermal stability properties of bleached, AN and MAN modified jute fibres. Each of the figure represents two thermogram curves, namely TGA and DTG. From Fig. 7a–c, it can be seen that the loss in weight is around 66.2% at 373.5°C, 55.7% at 384°C and 62.1% at 389°C for bleached, AN-modified and MAN-modified jute fibres, respectively. The main peak at about 355°C was due to the decomposition of α-cellulose (Fig. 7a). From the DTG curve in Fig. 7a–c, the rate of decomposition of bleached jute fibre was higher than that of AN- and MAN-modified jute fibres. Thus, the thermal stability of AN- and MAN-modified jute fibres are as high as 364 and 357°C, respectively. These are higher than that of the unmodified fibres. This can be explained by the incorporation of nitrile monomer on to the jute fibre.

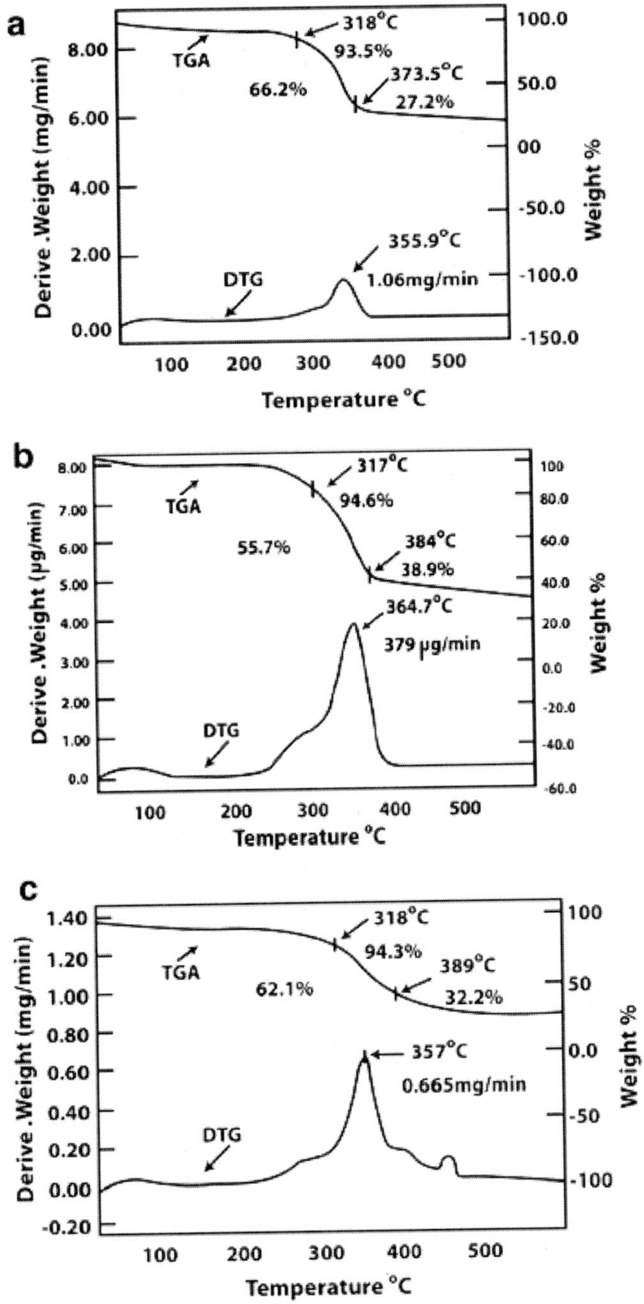

Figure 7. TGA and DTG of **a** bleached jute fibre, **b** acrylonitrile grafted jute fibre and **c** methacrylonitrile grafted jute fibre.

3.6 Surface morphology

Figure 8a–c shows the SEM micrograph of unmodified, AN-modified and MAN-modified jute fibres, respectively. The untreated cotton fibre shows the presence of a large number of micro-pores on its surface. After AN and MAN treatment, the jute fibre surfaces are coated with an outer layer of nitrile monomer (Rashidi et al. 2004) as observed in Fig. 8a, b. The ruptured surface of the modified fibre indicates the excess deposition of the nitrile monomer on the fibre hair, present in the jute fibre.

3.7 Determination of breaking strength of the bleached and modified jute fibres

It can be observed from Table 2 that the breaking load of modified jute fibre is higher that of the bleached fibre. It is also seen that the breaking load of MAN-modified fibre is higher (19.20 kg/Yarn) than that of AN-modified fibre (17.5 kg/Yarn). The difference may be due to the fact that poly-methacrylonitrile is a rigid polymer and poly-acrylonitrile is a soft polymer (Mondal et al. 2004).

4. CONCLUSIONS

In the present work, we have presented the graft copolymerization of nitrile monomers on to bleached jute fibre was carried out in the presence of a redox system, in an aqueous medium. The effect of the graft yield of the nitrile monomers on the jute fibre depends on the parameter variables and on the jute fibre as well. The chemical attachment between the nitrile monomer and hydroxyl group of cotton fibres was evaluated by FTIR. The grafted fibre showed improved physicochemical properties like tensile strength, moisture absorption and thermal stability. Thus a new type of jute fibre was obtained through graft polymerization. Such a process can make jute more suitable, as an input, in the manufacture of garments, home textiles etc.

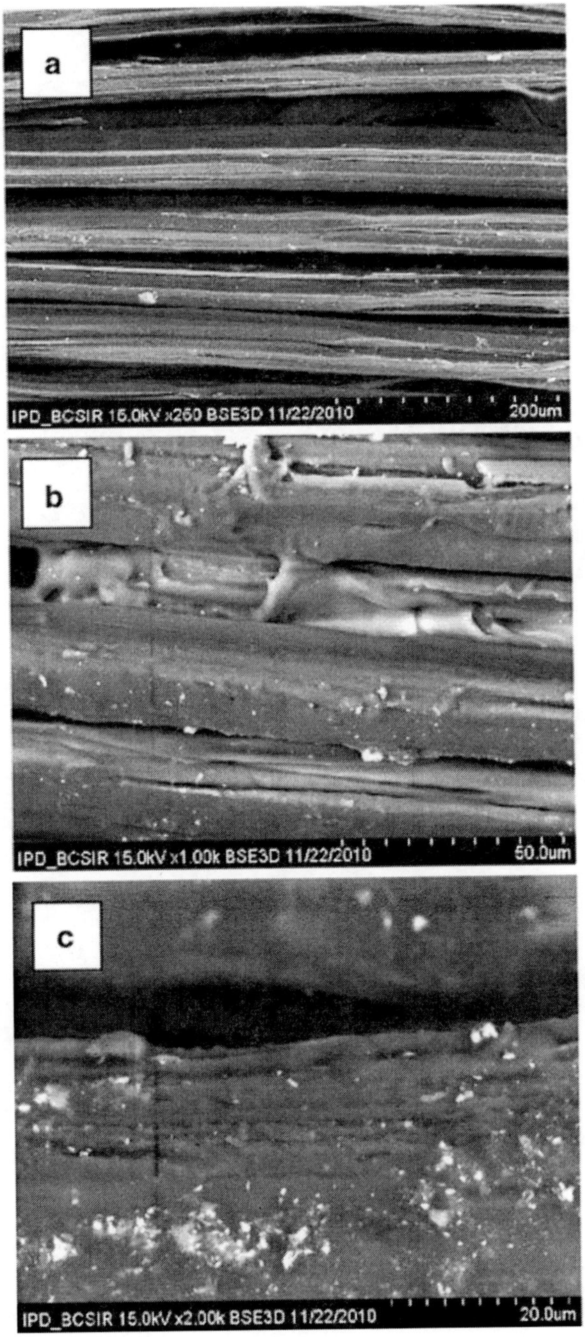

Figure 8. SEM of **a** bleached fibre, **b** acrylonitrile grafted jute fibre and **c** methacrylonitrile grafted jute fibre.

Table 2. Breaking strength of bleached and modified fibres

Bleached fibre	Modified fibre	
	AN-grafted fibre	MN-grafted fibre
Breaking load (kg/Yarn)		
16.02 ± 0.42	17.5 ± 0.39	19.20 ± 0.56

Results are expressed as mean ± standard deviation; n = 4.

AUTHORS' CONTRIBUTIONS

IHM designed the paper, performed the literature survey and completed the paper. KI performed the experiments and analyzed the data in consultation with IHM. All authors read and approved the final manuscript.

ACKNOWLEDGEMENTS

The authors would like to acknowledge the Ministry of Education in Bangladesh for funding the project as Higher Education Research Grant in 2014 (Memo no.: 37.01.0000.078.02.018.13-206(38)/6-35, 2014).

REFERENCES

1. Alam, M. S., & Khan, G. M. A. (2007). Chemical analysis of okra bast fiber (*Abelmoschus esculentus*) and its physicochemical properties. *Journal of Textile and Apparel Technology and Management, 5*(4), 1–9.

2. Bakr, N. A., Chee, C. Y., Abdullah, L. C., Ratnam, C. T., & Ibrahim, A. (2015). Thermal and dynamic mechanical properties of grafted kenaf filled poly(vinyl chloride)/ethylene vinyl acetate composites. *Materials and Design, 65*, 204–211.

3. Bianchi, E., Bonazza, A., Marsano, E., & Russo, S. (2000). Free radical grafting onto cellulose in homogeneous conditions. 2. Modified

cellulose–methyl methacrylate system. *Carbohydrate Polymer, 41*, 47–53.

4. Bicak, N., Sherrington, D. C., & Senkal, B. F. (1999). Graft copolymer of acrylamide onto cellulose as mercury selective sorbent. *Reactive Functional Polymer, 41*, 69–76.

5. Dong, A., Fan, X., Wang, Q., Yu, Y., & Cavaco-Paulo, A. (2015). Hydrophobic surface functionalization of lignocellulosic jute fabrics by enzymatic grafting of octadecylamine. *International Journal of Biological Macromolecules, 79*, 353–362.

6. El-Rafie, M. H., Abdul Hafiz, S. A., Hassan, S. M., & Hebeish, A. (1994). Grafting of methacrylic acid to loomstate viscose fabric using $KMnO_4/NaHSO_3$ system. *Polymers and Polymer Composites, 2*, 99–104.

7. El-Shekeil, Y. A., Sapuan, S. M., Jawaid, M., & Al-Shuja'a, O. M. (2014). Influence of fiber content on mechanical, morphological and thermal properties of kenaf fibers reinforced poly(vinyl chloride)/thermoplastic polyurethane poly-blend composites. *Materials and Design, 58*, 130–135.

8. Farouqui, F. I., & Mondal, Md I H. (1989). *Scouring and bleaching of jute fiber in relation to its strength* (pp. 1–8). XVII: The Rajshahi University Studies Part-B.

9. Fatimah, A. B. N., Ching, Y. C., Chuah, A. L., Thevy, R. C., & Azowa, I. N. (2014). Effect of methyl methacrylate grafted kenaf on mechanical properties of polyvinyl chloride/ethylene vinyl acetate composites.*Composites: Part A, 6*, 45–50.

10. Gonaslves, K. E., & Mungara, P. M. (1996). Synthesis and properties of degradable polyamides and related polymers. *Trends in Polymer Science, 4*(1), 25–31.

11. Hebeish, A., & Mehata, P. C. (1968). Cerium-initiated grafting of acrylonitrile on to cellulosic material. *Journal of Applied Polymer Science, 12*, 625–1647.

12. Hebiesh, A., Katouch, A., & EL-Rafie, M. H. (1971). Graft co-polymerization of vinyl monomer with modified cotton. II Grafting of acrylonitrile and methyl methacrylate on acetylated cotton. *Journal of Applied Polymer Science, 15*, 1–24.

13. Huque, M. M., Habibuddowla, M. D., Mahmood, A. J., & Mian, A. J. (1980). Graft-copolymerization onto jute fiber—ceric ion-initiated graft-copolymerization of methyl-methacrylate. *Journal of Polymer Science Part A Polymer Chemistry Edition, 18*, 1447–1458.

14. International Standard ISO 5081–1977(E). (1997). *Determination of breaking strength and elongation (Strip method)*. Switzerland: International Organisation for Standardization.

15. Khullar, R., Varshney, V. K., Naithani, S., & Soni, P. L. (2008). Grafting of acrylonitrile on to cellulosic material derived from bamboo (*Dendrocalamus strictus*). *Express Polymer Letters, 2*, 12–18.

16. Mcdowall, D. J., Gupta, B. S., & Stannett, V. T. (1984). Grafting of vinyl monomers to cellulose by ceric ion inhibition. *Progress of Polymer Sciences, 10*, 1–50.

17. Mishra, S., Misra, S. M., Tripathy, S. S., Nayak, S. K., & Mohanty, A. K. (2001). Graft copolymerization of acrylonitrile on chemically modified sisal fibres. *Macromolecular Materials and Engineering, 286*(2), 107–113.

18. Misra, M., Mohanty, A. K., & Singh, B. C. (1987). A study on grafting of methyl methacrylate on to jute fibre ($S_2O_8{}^{2-}$ thio system). *Journal of Applied Polymer Science, 33*, 2809–2819.

19. Mohanty, A. K., Khan, M. A., & Hinrichsen, G. (2000). Influence of chemical surface modification on the properties of biodegradable jute fabrics-polyester amide composites. *Composites: Part A, 31*, 143–150.

20. Mohanty, A. K., Singh, B. C., & Mirsha, B. C. (1986). Vanadium-cyclohexane-initiated graft copolymerization of methyl methacrylate on to jute fibre. *Journal of Applied Polymer Science, 31*, 1763–1769.

21. Mondal, Md I H. (2013). Grafting of methyl acrylate and methyl methacrylate on to jute fiber: physico-chemical characteristics of the grafted jute. *Journal of Engineered Fibers and Fabrics, 8*(3), 42–50.

22. Mondal, Md I H, Alam, R., & Sayeed, M. A. (2004). Graft copolymerization of nitrile monomers on to bleached jute fibre using potassium persulfate system and their textile characteristics. *Journal of Applied Polymer Science, 92*, 3622–3629.

23. Mukhopadhyay, S., Prasad, J., & Chatterjee, S. R. (1975). Grafting of acrylic acid on to methylcellulose. *Die Macromoleculare Chemie, 176*, 1–7.

24. Okieimen, E. F. (1987). Studies on the graft copolymerization of cellulosic materials. *European Polymer Journal, 23*, 319–322.

25. Ouajai, S., Hodzic, A., & Shanks, R. A. (2004). Morphological and grafting modification of natural cellulose fibers. *Journal of Applied Polymer Science, 94*, 2456–2465.

26. Ouajai, S., Hodzic, A., & Shanks, R. A. (2007). Morphological and grafting modification of natural cellulose fibras. *Materials Letters, 61*, 2456–2465.

27. Patra, C. M., & Singh, B. C. (1994). Influence of *N*-acetylglycine on the kinetics of ceric ion-initiated graft-copolymerization of acrylonitrile and methyl-methacrylate on to jute fibres. *Journal of Applied Polymer Science,52*, 1557–1568.

28. Rahman, M. V. (1970). Thermogravimetric and differential thermal analysis of cellulose, hemicelluloses and lignin. *Journal of Applied Polymer Science, 14*, 1323–1337.

29. Rashidi, A., Moussavipourgharbi, H., & Mirjalili, M. (2004). Effect of low-temperature plasma treatment modification of cotton and polyester fabrics. *Indian Journal of Fibre & Textile Research, 29*, 74–78.

30. Salam, M. A. (2005). Graft copolymerization of methacrylonitrile monomer on to sulfonated jute cotton-blended fibre. *Journal of Textile and Apparel, Technology and Management, 4*, 45–56.

31. Samal, R. K., Dash, S., & Swain, A. K. (1989). Graft copolymerization of methyl acrylamide on to jute fibers initiated by peroxydisulfate catalysed by Fe(III). *Polymer Journal, 21*, 821–828.

32. Samal, B. B., Sahu, S., Chinara, B. B., Nanda, S., Otta, P. K., Mohapatro, L. M., et al. (1988). Grafting of vinyl monomers onto sisal fiber in aqueous solution. *Journal of Polymer Science Part A Polymer Chemistry Edition,26*, 3159–3166.

33. Samal, R. K., Samantaray, H. S., & Samal, R. N. (1986). Graft copolymerization with a new class of acidic peroxo salt IV. Grafting of

acrylamide on to jute fibre using potassium mono persulphate: catalyzed by Fe(II).*Polymer Journal, 18*, 471–478.

34. Tsuboi, M. (1957). Infrared spectrum and crystal structure of cellulose. *Journal of Polymer Science, 2*, 157–171

Chapter 6

Degradation and detoxification of synthetic dyes and textile industry effluents by newly isolated *Leptosphaerulina* sp. from Colombia

Jersson Plácido, Xiomara Chanagá, Santiago Ortiz-Monsalve, María Yepes and Amanda Mora

Producción, Estructura y Aplicación de Biomoléculas, Universidad Nacional de Colombia Sede Medellín, Medellín, Colombia

ABSTRACT

Background

Wastewaters from the textile industry are an environmental problem for the well-known Colombian textile industry. Ligninolytic fungi and their enzymes are an option for the treatment of these wastewaters; however, the Colombian biodiversity has not been deeply evaluated for fungal strains with ligninolytic activities. In this research, 92 Colombian fungal isolates were collected from four locations around the Aburrá valley, Antioquia, Colombia. Their decolorizing activities were evaluated using Novacron Red,

Remazol Black and Turquoise Blue in solid and liquid media at different culture conditions. The best fungal isolate was evaluated in the bioremediation of two real effluents and its enzymatic extracts were used in the decolorization of the three dyes.

Results

From 92 Colombian fungal isolates, *Leptosphaerulina* sp. exhibited the best decolorization percentage (>90 %) in solid and liquid cultures, and in agitated and un-agitated conditions. *Leptosphaerulina* sp. effectively decolorized the three dyes and two real effluents from textile industries. This decolorization was catalyzed by the production of significant quantities of laccase (650 U/L) and manganese peroxidase (100 U/L). *Leptosphaerulina* sp. enzymatic extracts exhibited decolorizing activity when ABTS as mediator was added. *Leptosphaerulina* sp. decolorized two real effluents from textile industries (>90 %) under conditions of low pH and glucose supplementation. Enzymatic degradation and decolorization products' innocuity was demonstrated by cytotoxic and chromatographic analyses.

Conclusion

Leptosphaerulina sp. was the best Colombian isolate. This fungal strain achieved a decolorization above 90 % for the three dyes and two real effluents from a textile industry. This decolorization was performed by producing significant amounts of laccase and manganese peroxidase. *Leptosphaerulina* sp. is an interesting prospect to treat waters polluted with dyes without the production of compounds dangerous for the environment.

KEYWORDS

Leptosphaerulina sp., Ligninolytic enzymes, Dyes, Decolorization, Textile effluents, Colombia

1. BACKGROUND

The Colombian textile industry has national and international acknowledgements by its quality and tradition. However, this industry faces different environmental problems, one of which occurs during the fabrics dyeing. Dyeing environmental problems are related to the large demand of water and the presence of remaining dyes in their effluents. In fact, 1 kg of dyed textile requires 100 L of water generating an effluent with a dye content between 5 and 50 % (Abadulla et al. 2000). These wastewaters are disposed in rivers and creeks all around the country, generating significant pollution. Chemical dyes pollute waters by changing its color and generating harmful effects to the aquatic ecosystems. Besides the anti-aesthetic effect produced in the effluents, the coloration reduces the solar light transmittance, affecting the aquatic organisms' photosynthetic activity (Michniewicz et al. 2008). Additionally, some dyes have exhibited mutagenic and carcinogenic effects (Weisburger 2002). The traditional technologies to degrade synthetic dyes (physical, physicochemical and biological) are not efficient enough. In these technologies, the dyes are transferred to other matrices (i.e., microbial cells) or the dyes degradation products become more toxic than the original dyes (Zouari-Mechichi et al. 2006).

An environmentally friendly strategy to degrade textile dyes is the application of ligninolytic fungi and their enzymes. Ligninolytic fungi developed an enzymatic system composed by three principal enzymes: laccase (E.C 1.10.3.2), lignin peroxidase (LiP) (E.C 1.11.1.14), and manganese peroxidase (MnP) (E.C. 1.11.1.13). Laccase or benzenediol: oxygen oxidoreductase belongs to the oxidoreductases class (Dias et al. 2007). The most referenced and well-known laccases belong to the wood-decay fungi or white-rot fungi. Wood decay fungi utilize these enzymes to degrade the recalcitrant structure that constitutes lignin and facilitate the access to hemicellulose and cellulose. Laccases have the advantages of being extracellular, inducible, low substrate specificity and an inexpensive cofactor (oxygen) (Zouari-Mechichi et al. 2006; Sergio 2006; Couto and Toca-Herrera 2006). On the other hand, LiP and MnP belong to the peroxidases subclass and can catalyze the degradation of a wide number of aromatic structures, they are produced by the white rot fungi to participate in the initial states of the lignin degradation (MnP) (Angel2002; Hammel and Cullen 2008) or the degradation of aromatic rings moderately activated (LiP) (Khindaria et al. 1996). Therefore, ligninolytic fungi and their enzymes have been used in the degradation of different types of compounds as lignin, aromatic compounds, textile dyes, etc. (dos Santos et al. 2007).

Several fungi with ligninolytic activities (i.e., *Bjerkandera adusta, Trametes versicolor, Phanerochaete chrysosporium*) have proved their efficiency for treating textile dyes (dos Santos et al. 2007). However, each isolate generates a different enzymatic profile, a different response to environmental variables and different levels of dyes decolorization (Nyanhongo et al. 2002). Agitation, carbon and nitrogen source, and inducers (aromatic compounds, copper) are environmental factors that have been associated with the modification of the enzymatic production and/or the decolorization activity. However, the positive or negative effect of these factors is greatly related with the type of fungal strain evaluated. An example of this differences is the nitrogen concentration, low-nitrogen concentration has been found as favorable for*Cyathus bulleri* and *Phlebia tremellosa* (Salony et al. 2006; Robinson et al. 2001) whereas, *Pestalotiopsis*sp. and *Bjekandera adusta* showed greater enzymatic and decolorizing activity when the culture medium had nitrogen availability (Robinson et al. 2001; Hao et al. 2007).

In addition to fungal decolorization, synthetic dyes can be decolorized by using enzymatic extracts, purified or semi-purified enzymes, and commercial or native enzymes (Mattinen et al. 2011). Enzymatic decolorization have been performed by using individual or a mixture of ligninolytic enzymes, generally laccase is the main enzyme followed by MnP and LiP (Costa et al. 2005; Archibald et al. 1997). The principal advantage of enzymatic processes is the ability to obtain the same removal than microbial pretreatment in less time (between 2 and 48 h). However, some enzymes are not able to produce the same results when in vivo than in vitro. Therefore, some approaches have been utilized to increase the range of dyes degraded, the decolorization percentage, and the removal time (Zouari-Mechichi et al.2006; Kokol et al. 2007). The principal strategy is the use of mediators. Mediators are chemical compounds, which act as electron carriers between the enzyme and the final substrate (Babot et al.2011). The use of fungal laccases in presence of redox mediators is known as laccase-mediator system (LMS). The two most common mediators are 1-hydroxybenzotriazole (HBT) and 2,2'-azino-bis(3-ethylbenzothiazoline-6-sulphonic acid) (ABTS). A greater degree of degradation using LMS is based in a high level of interaction between the enzyme and its mediator; in fact, not all the mediators produce the same effect with enzymes from different origins (Nyanhongo et al. 2002).

Enzymatic or fungal processes using ligninolytic fungi and their enzymes are a promising technology to solve the dyes pollution generated by the Colombian textile industry. However, the fungal strains evaluated in the treatment of textiles dyes have been isolated from environmental and geographical conditions different than Colombia. Therefore, to apply this type of technology in Colombia, it is necessary to evaluate the Colombian microflora and their ligninolytic and decolorizing activities. Therefore, this research aims were the isolation of native fungi from four locations around the Aburrá valley, in Antioquia, Colombia, the evaluation of fungal strains effectiveness over synthetic dyes and actual effluents from a Colombian textile industry, and the evaluation of enzymatic extracts in the removal of synthetic dyes.

2. METHODS

2.1 Fungal strains

Ninety two fungi were isolated from four locations (cloud forest (CF), Montane rainforest (MRF), Medellin's Botanic Garden (BG) and an industrial and residential zone (IRZ)) around the Aburrá valley, Antioquia, Colombia. The strains were conserved in potato dextrose agar (PDA) at 4 °C until use.

2.2 Textile dyes

The dyes evaluated in this research were Novacron Red (azoic dye, λ_{max} = 514 nm), Remazol Black (azoic dye, λ_{max} = 598 nm), and Turquoise Blue (phthalocyanine dye, λ_{max} = 622 nm and 669 nm). Additionally, two types of real textile wastewaters were used. First, a dyeing bath effluent and second, the final effluent released into Medellin river. All dyes and effluents were kindly donated by Fabricato-Tejicondor S.A.

2.3 Evaluation of ligninolytic and decolorizing activity in solid media

The ligninolytic activity was measured in petri dishes with malt extract medium supplemented with 2,2′-azino-bis (3-ethylbenzothiazoline-6-sulphonic acid) also known as ABTS. The petri dish was inoculated with the isolated fungus and was incubated 2 weeks at 30 °C. The enzymatic activity was evaluated by the formation of a green or violet halo. The fungal strains that exhibited ligninolytic activity were selected for the decolorizing activity experiment. The decolorizing trials used malt extract agar supplemented with each dye. The fungi were cultivated during 20 days at 28 °C. The decolorizing activity was described as positive when the solid media exhibited partial or total loss of color. The fungal strains with decolorizing activity were selected for the liquid media evaluation.

2.4 Selection of decolorizing fungal strains

The fungal strains selection was performed in two phases. The first phase employed a completely randomized block design with three replicates. All the strains selected in the solid media experiment were evaluated in each dye. The culture was carried out in Erlenmeyer flasks with 50 mL of malt extract liquid media, 200 ppm of dye and an inoculum of four cylinders from the fungus. The fungal strains were cultured during 15 days at room temperature (20–25 °C) without agitation. At the end of the 15 days, the fungi with the largest decolorization percentage in each dye were selected for the next phase. The second phase of the selection process utilized the fungal strains selected in the unagitated culture, Kirk's media and agitation. The experiment employed a factorial design with three replicates. The experiment's factors were the selected fungi and ammonium tartrate concentration (0.22, 5, and 10 g/L). The fungal strains were cultivated during 15 days at room temperature and 150 rpm. The experiment's statistical analysis employed SAS 9.3 software. After that, a decolorization kinetic was performed for the treatments that exhibited the largest decolorization in each dye. The kinetic described the decolorization percentage, enzymatic activity, biomass, pH, protein concentration, glucose consumption, and adsorption. In order to determine the decolorization kinetic, Kirk's media was employed, with 200 ppm of dye, during 21 days at 25 °C and 150 rpm.

2.5 Selected fungus identification

The fungal isolate that presented the largest decolorization was molecular identified by analyzing the ITS and DNAr 28S sequences. The PCR used the following primers: ITS1 (5′ TCC GTA GGT GAA CCT GCG G 3′) and ITS4 (5′ TCC TCC GCT TAT TGA TAT GC 3′) and DNAr 28S LROR (5′ ACC CGC TGA ACT TAA GC 3′) and LR6 (5′ CGC CAG TTC TGC TTA CC 3′) (Chanagá Vera et al. 2012). The sequences analysis was performed in the BLAST software.

2.6 Enzymatic extracts experiment

The enzymatic extracts from the fungi with the largest enzymatic activity were evaluated in order to know their decolorizing activity. The cultures using the best fungus were stopped on the day with the largest enzymatic production. On this day, the media was centrifuged 10 min at 5000 rpm, after which, the supernatants were micro-filtrated and lyophilized. The enzymatic extracts experiment was performed for each dye, using a mixture of lyophilized extract (50 ppm), ABTS (2 mM), dye (50 ppm) and buffer pH 3. The enzymatic reaction was kept for 24 h at 25 °C and 150 rpm. The experiment had three replicates and the decolorization percentage as response variable. The statistical analysis was performed using the SAS 9.3 software.

2.7 Fungal decolorization of real textile wastewaters

The fungus with the greatest decolorization ability was evaluated with two types of textile wastewaters (the final effluent disposed into Medellin river and a fabrics dye bath). The final effluent was evaluated under a completely randomized factorial design 2 × 2 × 2 with three replicates and using the decolorization percentage as response variable. The experiment's factors were medium supplementation (10 g/L of glucose and 5 g/L of ammonium tartrate), pH (as received and pH 5) and inoculum (4 and 8 cylinders). The fabrics dyeing bath was evaluated under a completely randomized factorial design 2 × 2 × 2, using as factors medium supplementation (10 g/L of glucose and 5 g/L of ammonium tartrate), inoculum (4 and 8 cylinders) and effluent dilution [as received (6.8 g/L) and 200 ppm]. The decolorization was carried out in 250 mL Erlenmeyer flasks with 50 mL of medium, during 21 days at 25 °C and 150 rpm.

2.8 Enzymatic activities

The laccase activity was calculated using the enzymatic oxidation of ABTS. The reaction was followed spectrophotometrically by employing the change in the absorbance at 420 nm and a molar extinction coefficient of ε_{420}, 36,000/M/cm. The LiP activity was obtained by following the formation of veratryl aldehyde from veratryl alcohol at 310 nm (ε_{310}, 9300/M/cm). The MnP activity was calculated by measuring the absorbance change at 469 nm produced by the oxidation of 2,6-dimethoxypenol (DMF) (ε_{469}, 27,500/M/cm). All the enzymatic activities were reported in units (U), which were defined as the quantity of enzyme that catalyze 1 µmol of substrate per minute.

2.9 Analytical methods

The decolorizing activity was calculated using the decolorization percentage decolorization%=A0−AfA0×100decolorization%=A0−AfA0×100, where A_0 is the initial dye absorbance at the dye's λ_{max} and A_f is the final absorbance at the dye's λ_{max} (Mohorčič et al. 2006). The decolorizing activity was considered high when the decolorization percentage was above 80 %, medium when it was between 50 and 80 % and low when it was below 50 %. The error bars utilized in the decolorization percentage measurements corresponds to the standard error. To evaluate biomass adsorption, the biomass was re-suspended in methanol for 30 min, and the methanol solution was used to calculate the absorbance at the dye λ_{max}. In order to obtain the biomass production, after the methanol extraction, the biomass was dried at 105 °C for 24 h until obtain constant weight. The glucose determination was performed using the 3,5-dinitrosalicylic acid (DNS) methodology. The protein concentration was obtained by the Bradford's method. Dyes' degradation products were analyzed by thin layer chromatography (TLC) and 3-(4,5-Dimethylthiazol-2-yl)-2,5-diphenyltetrazolium bromide (MTT) cytotoxicity test. Lyophilized extracts from fungal cultures were employed for both analyses. In the TLC, the lyophilized extracts were solubilized in different organic solvents (methanol, acetonitrile, ethyl acetate, and hexane). Each organic extract was spotted in the solid phase (silica gel U60 plates) and its respective extraction solvent was used as mobile phase. The degradation-products toxicity was evaluated using the MTT test. This test evaluated the effect of the lyophilized extracts in the cell-viability percentage of human cell line U937 (Gallego et al. 2006). The cytotoxicity was measured by comparing the absorbance (595 nm) produced by the control cells and the absorbance produced by the cells exposed to different concentration of the lyophilized extracts (500, 250, 125, 62.5, 31.2, 15.6, and 7.81 ppm). A compound was considered toxic, when the absorbance of the cells exposed to the compound exhibited an absorbance lower than the control's absorbance (Gallego et al. 2006).

3. RESULTS

3.1 Evaluation of ligninolytic and decolorizing activity in solid media

In total, 92 fungi were collected from the four isolation locations. The strains were obtained from different types of lignocellulosic material (composted biomass, trees, discomposing trees etc.). The isolates distribution by site was IRZ (15), BG (32), CF (25), and MRF (20). After isolation and strains purification, the fungi were inoculated in solid malt extract media supplemented with ABTS. In solid media was possible to observe green and purple halos and a combination of both halos. The purple halos were observed after the production of small green halos. The halo formation fluctuated from hours (12–24 h) to days (2–12 days). In total, 30 of 92 isolates exhibited ligninolytic activity. The isolates ID and isolation location are described in Table 1. The total percentage of fungi with enzymatic activity was 32 % and the percentages of positive fungi by sites were IRZ (60 %), BG (15.6 %), CF (40 %), and MRF (32 %).

The 30 selected isolates were evaluated in malt extract agar with each of the three dyes (Novacron red, Remazol black, and Turquoise blue). In this case, an isolate was classified as positive for decolorizing activity when the media exhibited partial or total decolorization. In total, 15 fungi exhibited decolorizing activity which corresponds to 50 % of the preselected fungi and 16.3 % of the original isolated fungi. The percentage of positive isolates was relatively high. In fact, other reports with more than 50 isolates obtained 5.7 % (Hernández-Luna et al. 2008) and 10 % (Risna and Suhirman 2002) of fungi with decolorizing activity. The fungal isolates with decolorizing activity and their level of decolorization are reported in Table 2. Remazol black was degraded by 12 strains, two of which (LVN6 and LVN63) showed the greatest decolorization. Additionally, LVN6 and LVN63 obtained the fastest decolorization with 5 and 3 days, respectively. Similar decolorization velocity have been reported in *P. chrysosporium* (Máximo et al. 2003), *T. versicolor* (Mohorčič et al. 2006), *Bjerkandera adusta* and *T. versicolor* (Novotný et al. 2004). In contrast, the other fungal strains took from 8 to 15 days in decolorize Remazol Black. On the other hand, Turquoise blue obtained the largest number of fungal strains with decolorization (13 strains); however, the decolorization realized by the fungal strains had medium or low level (Table 2). In Turquoise Blue, the greatest decolorization was obtained after day 15. Opposite to Turquoise Blue, Novacron Red obtained the lowest number of fungal isolates (4) with decolorizing activity. This low quantity of positive strains was related to the dye's monoazoic structure. Monoazoic dyes have been reported in different studies to exhibit a high difficulty to be degraded by ligninolytic fungi. This difficulty is associated with the enzyme access to the dye's azo group (–N = N–) (Máximo et al. 2003; Park et al. 2007).

Table 1. Fungal isolate and isolation location of the fungal strains with ligninolytic activity

Isolate	Isolation location	Isolate ID	Isolation location
LVN 4	Botanic garden	LVN 63	Cloud forest
LVN 5	Botanic garden	LVN 65	Cloud forest
LVN 6	Botanic garden	LVN 71	Cloud forest
LVN 31	Botanic garden	LVN 72	Cloud forest
LVN 32	Botanic garden	LVN 73	Cloud forest
LVN 35	Montane rainforest	LVN 74	Cloud forest
LVN 42	Montane rainforest	LVN 79	Industrial and residential zone
LVN 44	Montane rainforest	LVN 80	Industrial and residential zone
LVN 45	Montane rainforest	LVN 82	Industrial and residential zone
LVN 48	Montane rainforest	LVN 83	Industrial and residential zone
LVN 49	Montane rainforest	LVN 84	Industrial and residential zone
LVN 51	Montane rainforest	LVN 85	Industrial and residential zone
LVN 52	Montane rainforest	LVN 88	Industrial and residential zone
LVN 54	Cloud forest	LVN 91	Industrial and residential zone
LVN 61	Cloud forest	LVN 92	Industrial and residential zone

Table 2. Qualitative evaluation of the decolorizing ability in the selected fungal strains

Isolate	Dye		
	Turquoise Blue	Remazol Black	Novacron Red
LVN 6	Medium	High	Medium
LVN 35	Low	Low	Absence
LVN 44	Low	Low	Absence
LVN 45	Low	Low	Absence
LVN 48	Low	Absence	Absence
LVN 49	Medium	Absence	Absence
LVN 52	Absence	Medium	Absence
LVN 54	Medium	Low	Absence
LVN 61	Medium	Absence	Absence
LVN 63	Low	High	Medium
LVN 71	Low	Low	Absence
LVN 72	Low	Absence	Absence
LVN 74	Low	Medium	Absence
LVN 80	Low	Medium	Medium
LVN 88	Absence	Medium	Low

3.2 Selection of decolorizing fungal strains

The fungal strains selection was performed in two phases. The first phase was a preselection study employing malt extract medium without agitation. Whereas, the second phase was an experiment using agitation, Kirk's medium and variations in the nitrogen concentration.

3.3 Fungal strains preselection

The preselection experiment was carried out with the 15 isolates that exhibited decolorization activity in solid media (Fig. 1). This experiment evidenced that all the strains generated decolorization; however, the removal level varied among strains and dyes. The variation between solid and liquid media is explained by the changes in environmental conditions, it has been found that fungi behaves different between solid and submerged cultures, in fact, enzymes production is favored by growing in solid surfaces improving the enzymatic activity and the enzymatic profile (Novotný et al. 2004). In Remazol Black, the isolates LVN6, LVN88, and LVN63 exhibited the highest removal (>80 %), whereas the other fungi removal percentages varied between 20 and 80 %. The results between solid and liquid experiments were similar in LVN6 and LVN63; in both, they achieved the largest decolorization. In Novacron Red, the fungal isolates LVN80, LVN61, LVN6, LVN63 and LVN88 exhibited the most significant decolorization. From these fungi, LVN61 was the only strain with high decolorization in the liquid experiment and low decolorization in the solid experiment. The high removal observed in Novacron Red, under non-agitated conditions, is not common. In fact, monoazoic dyes had low or none decolorization under non-agitated conditions (Jarosz-Wilkołazka et al. 2002; Swamy and Ramsay 1999). Turning now to turquoise blue, five strains (LVN6, LVN49, LVN63, LVN80, and LVN72) obtained decolorization around 70 % or more. This result is promissory because under non-agitation conditions other phthalocyanines dyes have not been decolorized by other fungal strains (Swamy and Ramsay 1999).

The experiment's statistical analysis (see Additional file 1: Annex S1) was based in the decolorization percentage, using the Duncan's test to select the best fungal isolates for each dye (Table 3). The isolates selected were LVN6, LVN63, LVN80, LVN88, LVN61, LVN49, and LVN72. The decolorization percentage of the selected fungal strains were around 85 % (see Fig. 1). These values were similar to the values achieved by cultures without agitation of *Phlebia tremellosa, Irpex lacteus,* and basidiomycete strain PV 002 (Kirby et al. 2000; Svobodová et al. 2008). The isolates LVN6 and LVN63 were the only strains with high removal in all the dyes evaluated. The ability of decolorize different types of dyes, under non-agitation conditions, was reported in a small number of fungal strains (Kirby et al. 2000; Erkurt et al.2007).

3.4 Fungal strains selection

The fungal strains selection was performed in Kirk's medium with agitation and three different concentrations of ammonium tartrate. The fungal strains used in each dye are shown in Table 3. The experiment employed a factorial design, using as factors the selected microorganism for each dye and the three concentrations of ammonium tartrate. The factorial design produced nine treatments for Remazol Black and 15 treatments for Novacron Red and Turquoise Blue. The decolorization percentages obtained by each treatment are shown in Fig. 2.

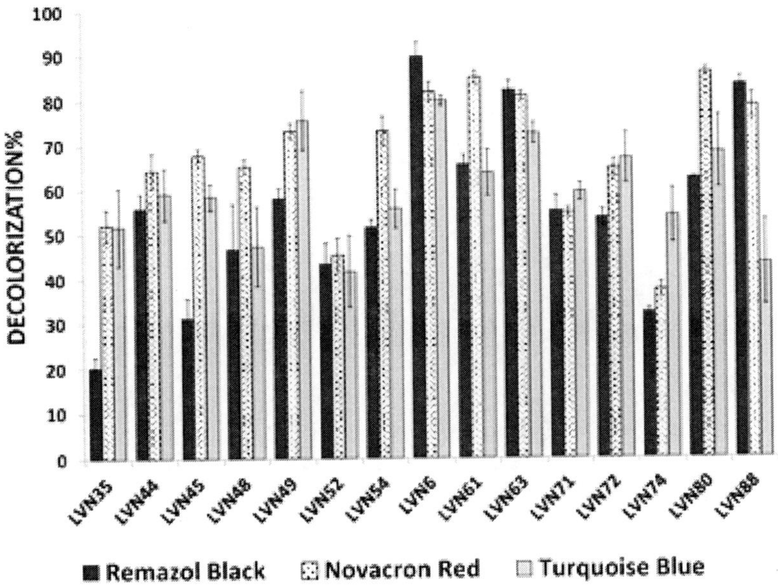

Figure 1. *Barplot* of the decolorization percentage obtained by the fungal strains in the preselection experiment without agitation and malt extract media

Table 3. Preselected fungal isolates

Turquoise Blue	Remazol Black	Novacron Red
LVN 6	LVN 6	LVN 6
LVN 63	LVN 63	LVN 63
LVN 80	LVN 88	LVN 80
LVN 72		LVN 88
LVN 49		LVN 61

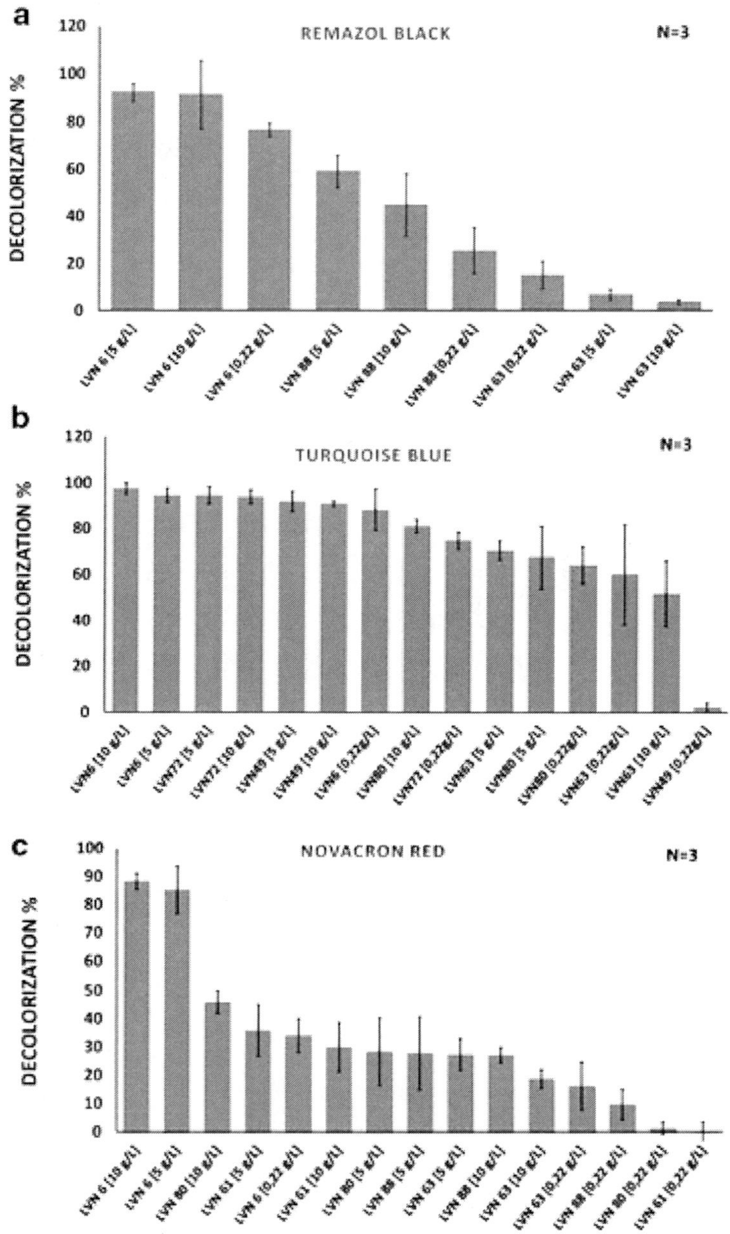

Figure 2. *Barplot* of the decolorization percentage obtained in the selection experiment with Kirk's media and agitation; **a** Remazol Black, **b** Turquoise Blue, **c** Novacron Red

Remazol Black experiments showed LVN6 as the best fungal strain (>90 %). In this dye, LVN6 at the three concentrations of ammonium tartrate realized decolorization greater than the other fungal strains (Fig. 2a). The statistical analysis showed interaction between the factors; therefore, the treatments selection was performed using the interaction plots (see Additional file 1: Annex S1). The interaction plots evidenced the greatest decolorization in the treatments with LVN6 and 5 and 10 g/L of ammonium tartrate. In contrast, LVN63 obtained the highest removal at 0.22 g/L of ammonium tartrate; this behavior was also detected in Novacron Red. The decolorizing activity observed in LVN6 [5 g/L] and LVN6 [10 g/L] was similar to *Bjerkandera adusta, Trametes versicolor, Funalia trogii* and *Irpex Lacteus,* strains recognized by their high decolorizing activity over Remazol Black (Mohorčič et al. 2006; Park et al. 2007; Máximo and Costa-Ferreira 2004). In Remazol Black, the selected treatment was LVN6 [5 g/L] because it produced the highest decolorization with the lowest consumption of ammonium tartrate.

The highest removals were achieved in Turquoise Blue. In this dye, almost all the treatments presented removals above 50 %. In this dye, three strains (LVN6, LVN72 and LVN49) obtained decolorizations above 90 % (Fig. 2b). The decolorization obtained by LVN6, LVN72 and LVN49 is comparable with the removal obtained by *P. simplicissimum* INCQS40211, *P. chrysosporium,* and *Ph. Tremellosa* over Turquoise Blue. In these strains, the removal was above 94 % (Kirby et al. 2000; Bergsten-Torralba et al.2009). Similar to Remazol black, the statistical analysis evidenced the interaction between isolates and ammonium tartrate concentration. The interaction plots indicated LVN49, LVN72 and LVN6 as the best fungal strains to decolorize Turquoise Blue and the concentrations of 5 and 10 g/L of ammonium tartrate as the best concentrations to obtain high decolorization. In this dye, the best treatments were LVN49 [5 g/L], LVN72 [5 g/L], and LVN6 [5 g/L]. However, the treatments LVN49 [5 g/L] and LVN72 [5 g/L] were not employed in further experiments because its lack of efficiency over the other two dyes.

Novacron Red experiment exhibited LVN6 [5 g/L] and LVN6 [10 g/L] as the treatments with the greatest decolorization percentages (>85 %) (Fig. 2c). On contrary, the other fungal strains had a reduction in the decolorizing ability, in comparison with the unagitated experiment. This reduction can be explained by the changes in the growth media and the inclusion of agitation. The statistical analysis (see Additional file 1: Annex S1) exhibited significant differences among the treatments and an interaction between the factors. The interaction plots revealed 10 and 5 g/L of ammonium tartrate and LVN6 as the best treatments for Novacron Red decolorization. The removal obtained by LVN6 [5 g/L] (85.3 %) and LVN6 [10 g/L] (88.5 %) was larger than the reported by other fungal strains at similar or greater concentrations (≥200 ppm) of monoazoic dyes (Jarosz-Wilkołazka et al. 2002). The treatment LVN6 [5 g/L] was selected as the best option for the decolorization of Novacron Red, because it reached a similar decolorization than LVN6 [10 g/L], using half of the concentration of ammonium tartrate.

3.5 Molecular identification of the isolate LVN6

The ITS region (DNAr) and DNAr 28S from LVN6 were amplified to identify the molecular identity of this strain. The amplification of these DNA sequences generated bands of 500–600 pair of bases (pb) for ITS, and 1100 pb for DNAr 28S. The BLAST analysis and sequence identity of ITS and DNAr 28S identified the fungi LVN6 as a member of the genre *Leptosphaerulina*. A specie determination using the blast analysis was not possible. In this strain, the sequences of the ITS region and DNAr 28S were homologous to two different species of *Leptosphaerulina*. The ITS region's sequence was homologue to*Leptosphaerulina Americana*, whereas DNAr 28S' sequence was homologous to *L. australis.* The phylogenetic and morphological analysis of these sequences did not allow the complete identification of this strain (Chanagá Vera et al. 2012). At this moment, this is the first report demonstrating the decolorizing activity in the genre *Leptosphaerulina.* The ability of *Leptosphaerulina* sp. to decolorize solutions with dyes can be related to the presence of ligninolytic enzymes. Likewise, *L. chartarum* and *L. briosiana* have been described as ligninolytic enzymes producers (Simon et al. 1979; Sajben-Nagy et al.2014).

3.6 Decolorization kinetics

Figure 3a, b shown the Remazol Black decolorization kinetic. In the first 5 days, Remazol Black decolorization was approximately 30 %; however, after seven days, it increased above 90 %. The decolorization was correlated with the increment in the enzymatic production (Fig. 3a). *Leptosphaerulina*sp. produced laccase and MnP during all the kinetic. In Remazol Black, the greatest production of enzymes started at day 7 and had its maximum production around day 15. The highest production of laccase and MnP was 650 U/L and 100 U/L, respectively. Turquoise Blue decolorization kinetic exhibited a removal above 90 % in 3 days (Fig. 3c, d). The initial removal was performed by an adsorption process. The Turquoise Blue high adsorption can be related to the copper molecule in the core of the dye structure. After the day 5, the adsorption process was replaced by the enzymatic process. The enzymatic catalysis removed the dye particles attached to the fungal biomass. This removal was catalyzed by laccase and MnP production. The largest enzymatic production was observed at day 9 with a laccase activity of 512 U/L and a MnP activity of 100 U/L. The Novacron Red decolorization kinetic by*Leptosphaerulina* sp. [5 g/L] (Fig. 3e, f) showed a removal greater than 90 % and the production of ligninolytic enzymes. The highest decolorization achieved by *Leptosphaerulina* sp. was obtained in 9 days, the enzymatic production was observed during all the kinetic; nevertheless, the highest enzymatic activities of laccase and MnP were observed on the 30th day with 647 U/L and 90 U/L, respectively.

3.7 Enzymatic decolorization using *Leptosphaerulina* sp. extracts

The enzymatic degradation of the three dyes was evaluated by employing a commercial laccase and ABTS as mediator (see Additional file 1: Annex S1). The in vitro analysis showed that all the dyes were decolorized by the commercial laccase; however, the level of decolorization varied depending on the dye and ABTS concentrations. The highest decolorization was obtained in Turquoise Blue (83 %), followed by Remazol Black (80 %), and Novacron Red (56 %). The greatest decolorizations were achieved by the treatments with ABTS as mediator (2 mM) and the highest commercial enzyme concentration (2 mg/ml). Turquoise Blue was the only dye that exhibited any level of decolorization without using ABTS. The*Leptosphaerulina* sp. lyophilized-extracts (Fig. 4) decolorized enzymatically all the dyes. In Remazol Black, the enzymatic decolorization was approximately 70 %. *Leptosphaerulina* sp. enzymatic extract transformed the reaction media color, from a deep black into a reddish color. *Leptosphaerulina* sp. enzymatic extracts achieved similar levels of decolorization than the commercial enzyme (80 %). In Turquoise Blue, the *Leptosphaerulina* sp. enzymatic extracts produced a decolorization of 56 % (Fig. 4). In Turquoise Blue, the decolorization difference between the commercial laccase (83 %) and*Leptosphaerulina* sp. enzymatic extracts (56 %) evidenced a necessity to improve the enzyme purification process. In

Novacron Red, the enzymatic extracts achieved low removal (20 %); nevertheless, the commercial enzyme also got low removal (56 %).

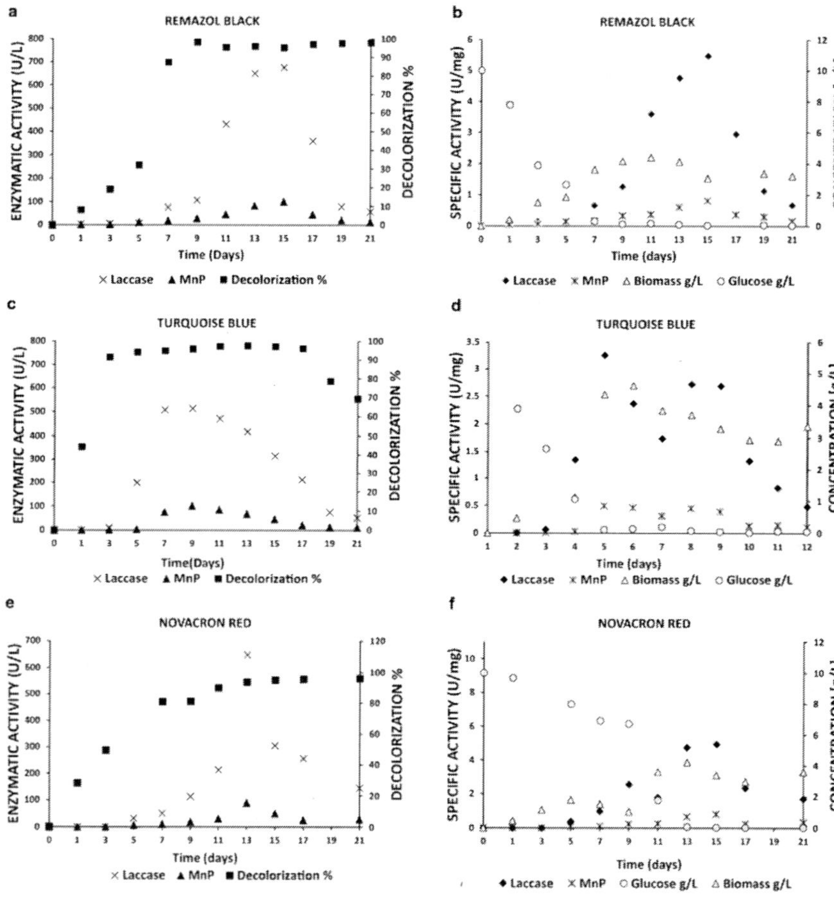

Figure 3. *Leptosphaerulina* sp. decolorization kinetic. **a** Remazol Black kinetic showing decolorization % and enzymatic activities, **b** Remazol Black kinetic showing glucose concentration, biomass concentration, and specific enzymatic activities **c**Turquoise Blue kinetic showing decolorization percentage and enzymatic activities, **d** Turquoise Blue kinetic showing kinetic showing glucose concentration, biomass concentration, and specific enzymatic activities **e** Novacron Red showing decolorization percentage and enzymatic activities, **f** Novacron Red showing showing kinetic showing glucose concentration, biomass concentration, and specific enzymatic activities

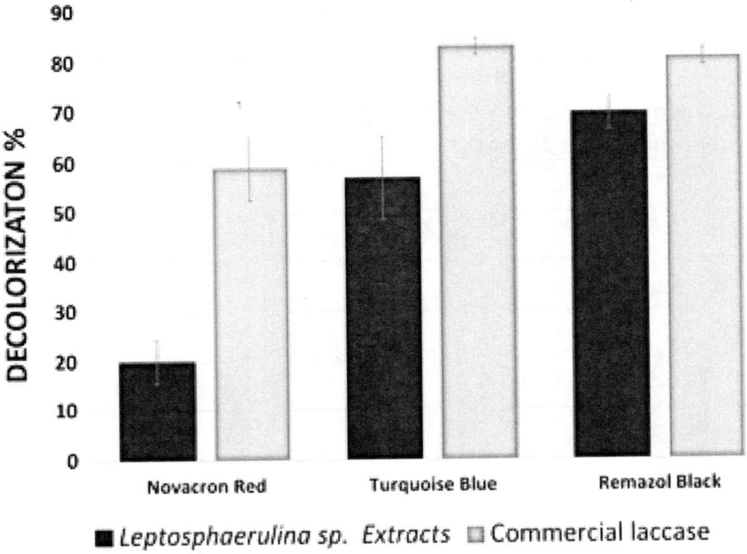

Figure 4. *Barplot* of the decolorization percentage obtained by the in vitro study of *Leptosphaerulina* sp. enzymatic extracts

3.8 Real textile effluents treatment using *Leptosphaerulina* sp.

The final discharge effluent had the following characteristics: alkaline pH (12), greyish color, low dyes concentration, and the presence of suspended particles. The effluent spectrophotometric analysis showed two strong signals at 520 and 680 nm. These signals were employed to calculate the effluent decolorization percentage. The decolorizing effect of *Leptosphaerulina* sp. was evaluated using a factorial design 2 × 2 × 2. The experiment's factors were: medium supplementation (10 g/L of glucose and 5 g/L of ammonium tartrate), pH (as received and pH 5) and inoculum (4 and 8 cylinders). After 21 days, the treatments with low pH achieved a decolorization above 70 %, whereas, the treatments with the original pH produced removals below 30 % (Fig. 5a). The experiment's statistical analysis evidenced a lack of interaction between the three factors (p value >0.05). Thus, the best treatment selection was carried out by doing the Duncan's test (see Additional file 1: Annex S1). The treatment pHm-Sup-8 achieved the best decolorization (95.5 %), followed by pHm-Sup-4 (71 %), pHm-woSup-4 (70 %), and pHm-woSup-8 (69 %). The treatments that produced the largest decolorization also showed laccase production. pHm-Sup-8 treatment achieved the greatest laccase production (24.86 U/L), followed by pHm-Sup-4 (5.27 U/L). The first and final days effluent's spectrum evidenced the complete reduction of the signals observed at the beginning of the experiment (Fig. 5c).

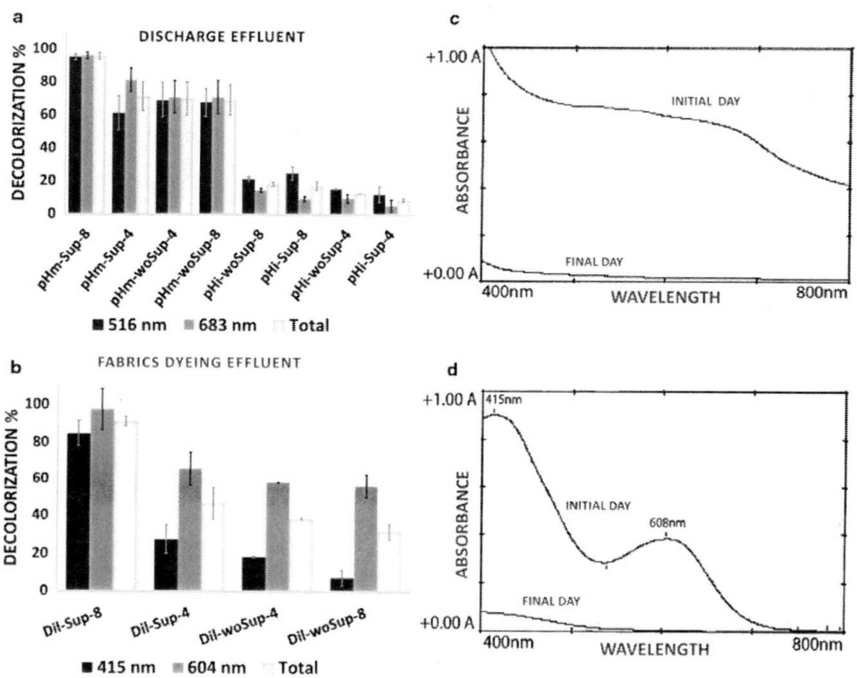

Figure 5. Real textile industry effluents decolorization by *Leptosphaerulina* sp. **a** *Barplot* of the discharge effluent experiment. **b** *Barplot* of the dyeing effluent experiment. **c** Discharge effluent spectra from pHm-Sup-8 treatment at the initial and final day. **d** Dyeing effluent spectra Dil-Sup-8 treatment at the initial and final day

The dyeing effluent was evaluated using a factorial design 2 × 2 × 2. In this effluent, the experiment's factors were: medium supplementation (10 g/L of glucose and 5 g/L of ammonium tartrate), inoculum (4 or 8 cylinders), and dilution (as received and 200 ppm of dyes, the dye concentration was calculated based in Novacron Red's original concentration). The effluent had three dyes: Novacron Red (6.8 g/L), Novacron Blue (6.75 g/L), and

Novacron Yellow (5.94 g/L). The effluent had a basic pH (12), suspended particles and deep military green coloration. The spectrophotometric analysis revealed two strong signals at 415 and 608 nm (Fig. 5d); these signals were used to calculate the decolorization percentage.*Leptosphaerulina* sp. only decolorized the effluent when it was diluted to 200 ppm (Fig. 5b), the undiluted samples did not show decolorization. The largest removal was obtained by Dil-Sup-8 (90 %), the other diluted treatments decolorized the effluent below 50 %. Dil-Sup-8 exhibited a considerable reduction in the signal at 415 nm (Fig. 5d). The absorbance reduction in this peak can be related with the production of laccases. In fact, Dil-Sup-8 presented the most significant laccase production (40 U/L). The absorbance reduction at 415 nm was observed in the treatment Dil-Sup-4, which also produced laccase activity (10 U/L). The statistical analysis evidenced an interaction between the inoculum quantity and the medium supplementation (see Additional file 1: Annex S1). The interactions plot described an inverse interaction between the factors. Additionally, the interaction plot confirmed the treatment Dil-Sup-8 as the best conditions to decolorize the dyeing effluent by *Leptosphaerulina* sp.

3.9 Decolorization products evaluation

The decolorized media (dyes and effluents) were lyophilized; then, the compounds in the lyophilized were extracted using four solvents (methanol, acetonitrile, ethyl acetate, and hexane). After that, the thin layer chromatography (TLC) was performed. The extracts were spotted and run it using the same solvent as mobile phase. The hexane extracts did not produce any type of signals under UV light (near and far) and visible light. By contrast, the methanol, acetonitrile and ethyl acetate extracts revealed signals under UV and visible lights (Table 4). The chromatograms of the three dyes and the effluents exhibited the disappearance of the signals corresponding to the dyes were present in dissolution and effluents (Rf 1). Additionally, the methanol, acetonitrile and ethyl acetate extracts showed a large number of signals. These signals were not observed in the controls (Kirk's media with and without dye). Turquoise Blue extracts generated a number of signals greater than Remazol Black and Novacron Red extracts. The use of TLC to evaluate the fungal or enzymatic degradation of dye have been reported in the enzymatic decolorization of Reactive Orange 16, Reactive Red 2 and Reactive Blue 4 (Axelsson et al. 2006; Telke et al. 2009). In these publications, the disappearance of the original dye's signal and the appearance of one or more signals with different Rf were correlated with fungal or enzymatic dye degradation. Using the same proposition, *Leptosphaerulina* sp. degraded the dyes in Kirk's media and the real effluents. However, it was not possible to identify how and which of the signals were related to the dye degradation.

The MTT test used lyophilized extracts from decolorized cultures of *Leptosphaerulina* sp. The MTT analyzed the cytotoxic effect of different concentrations of the lyophilized extracts over the human cell line U937. At the end of the test, the concentrations evaluated did not evidence cytotoxicity to the human cell line U937, because none of the concentrations generated a significant reduction in the absorbance at 595 nm (Table 5). These results evidenced that *Leptosphaerulina* sp. decolorized Novacron Red, Remazol Black and Turquoise Blue, without the production of toxic compounds.

4. DISCUSSION

From 92 fungi isolated from different ecosystems, the selection process performed in this paper indicated that *Leptosphaerulina* sp. is the fungal strain with the best decolorizing activity among the fungal strains isolated. At this moment, this is the first report demonstrating the decolorizing activity in the genre*Leptosphaerulina.* The ability of *Leptosphaerulina* sp. to decolorize solutions with dyes can be related to the presence of ligninolytic enzymes. Likewise, *L. chartarum* and *L. briosiana* have been described as ligninolytic enzymes producers (Simon et al. 1979; Sajben-Nagy et al. 2014).

The decolorizing activity observed by *Leptosphaerulina* sp. in remazol black was similar to *Bjerkandera adusta, Trametes versicolor, Funalia trogii* and *Irpex Lacteus,* strains recognized by their high decolorizing activity over Remazol Black (Mohorčič et al. 2006; Park et al. 2007; Máximo and Costa-Ferreira 2004). Similarly, *Leptosphaerulina* sp. decolorization percentage (>90 %) can be compared with the decolorization obtained by *P. simplicissimum* INCQS40211, *P. chrysosporium,* and *Ph. Tremellosa* over Turquoise Blue. In these fungal strains, the removal was above 94 % (Kirby et al. 2000; Bergsten-Torralba et al. 2009). Similarly, novacron red was decolorized by *Leptosphaerulina* sp. in a greater level than the reported by other fungal strains at similar or greater concentrations (≥200 ppm) of monoazoic dyes (Jarosz-Wilkołazka et al. 2002).

Table 4. Thin layer chromatography results of the lyophilized extracts obtained after the decolorization process

Dye/effluent	Solvent	Number of new bands			
		Visible	Near UV	Distant UV	Total
Remazol Black	Methanol	3 (Rf: 0.72, 0.78, 0.86)	3 (Rf: 0.72, 0.78, 0.86)	3 (Rf: 0.52, 0.72, 0.8)	9
	Acetonitrile	0	2 (Rf: band and 0.84)	2 (Rf: 1, 0.72)	4
	Ethyl acetate	0	1 (Rf: band between 0.70 and 0.79)	4 (Rf: 0.6,0.71, 0.79, 0.83)	5
Novacron red	Methanol	1 (Rf: 0.69)	3 (Rf: 0.67, band between 0.7 and 0.74, 0.79)	4 (Rf: 0.55, 0.75, 0.78, 0.87)	8
	Acetonitrile	0	3 (Rf: band, 0.52, 0.86)	3 (Rf: 0.68, 0.91, 1)	6
	Ethyl acetate	0	3 (Rf: 0.25, 0.31, 0.51, 0.75)	4 (Rf: 0.62, 0.79, 0.81, 0.85)	7
Turquoise blue	Methanol	1 (Rf: 0.69)	3 (Rf: 0.69, band between 0.71 and 0.77, 0.82)	4 (Rf: 0.55, 0.67, 0.7, 0.8)	8
	Acetonitrile	0	2 (Rf: barrido y 0.88)	5 (Rf: 0.76, 0.79, 0.83, 0.93, 1)	7
	Ethyl acetate	0	5 (Rf: 0.13, 0.19, 0.25, and 0.33 and a band between 0.66 and 0.8)	5 (Rf: 0.74, 0.78 0.81, 0.86, 0.9)	10
Discharge effluent	Methanol	0	2 (Rf: 0.65, 0.76)	1 (Rf: 0.65)	2
	Acetonitrile	0	1 (Rf: 1)	0	1
	Ethyl acetate	0	2 (band and Rf: 0.7)	6 (Rfs: 0.09, 0.16, 0.22, 0.5, 0.72, 0.84)	8
Dyeing effluent	Methanol	1 (Rf: 0.65)	1 (Rf: 0.65)	1 (Rf: 0.65)	1
	Acetonitrile	0	1 (Rf: 0.78)	0	1
	Ethyl acetate	0	3 (Rf: 0.65, 0.1, 0.2)	2 (Rf: 0.71, 0.76)	5

Table 5. Cytotoxicity analysis of the lyophilized extracts obtained after the decolorization process using the MTT test

Lyophilized extracts concentration (ppm)	Remazol Black Absorbance	Novacron Red Absorbance	Turquoise Blue Absorbance
500	1.09 ± 0.07	1.03 ± 0.04	1.02 ± 0.09
250	1.11 ± 0.10	0.68 ± 0.06	0.68 ± 0.10
125	0.99 ± 0.25	0.79 ± 0.05	0.68 ± 0.05
62.5	1.01 ± 0.05	0.81 ± 0.01	0.73 ± 0.07
31.2	1.04 ± 0.10	0.77 ± 0.02	0.68 ± 0.04
15.6	1.03 ± 0.05	0.64 ± 0.05	0.74 ± 0.08
7.8	1.15 ± 0.15	0.73 ± 0.09	0.63 ± 0.11
Control	0.91 ± 0.06	0.72 ± 0.06	0.72 ± 0.06

In terms of decolorization time, *Leptosphaerulina* sp. decolorized remazol black in times comparable with*Funalia trogii* (3 days); however, *F. Trogii* had a dye concentration (100 ppm) lower than *Leptosphaerulina*sp. (200 ppm) (Park et al. 2007). In other fungal strains, Remazol black decolorization was realized between 10 and 15 days and with a dye concentration lower than the used in this investigation (Mohorčič et al. 2006; Park et al. 2007). Turquoise Blue decolorization exhibited a high decolorization time (3 days) (Fig. 3b). However, in this dye, the initial removal was performed by an adsorption process. The Turquoise Blue high adsorption can be related to the copper molecule in the core of the dye structure. The adsorption of copper-related dyes has been observed in other fungal strains, as *P. simplicissimum*and *Penicillium oxalicum* (Bergsten-Torralba et al. 2009; Xin et al. 2010). In contrast, novacron red was decolorized in less days (9 days, 90 %); however, this period is considered as a fast decolorization for monoazoic dyes. Other fungal strains decolorized monoazoic dyes at the same concentration, during a time between 12–20 days (Swamy and Ramsay 1999; Toh et al. 2003). Faster decolorizations have been reported; nevertheless, these decolorizations used concentrations (80–100 ppm) less than this study (200 ppm) (Swamy and Ramsay 1999). Similar to turquoise blue, Novacron Red decolorization by*Leptosphaerulina* sp. exhibited two decolorization mechanisms, adsorption and enzymatic catalysis. Adsorption was the predominant mechanism during the 1st days; however, after the 3rd day, the principal mechanism was the enzymatic catalysis. The combination of these two removal mechanisms has also been described in *I. lacteus* and *Penicillium simplicissimum* in the decolorization of reactive red 198 and methyl red (Novotný et al. 2004; Bergsten-Torralba et al. 2009).

Leptosphaerulina sp. produced laccase and MnP to decolorize the three dyes, LiP was not observed in the cultures. It is important to remark that these enzymes were produced during all the culture; however, they reached their maximum production at different times. In remazol black, the largest enzyme production of laccase (650 U/L) and MnP(100 U/L) by *Leptosphaerulina* sp. was observed at day 15. Ligninolytic microorganisms that decolorize solutions with Remazol Black can be grouped in two sets, high laccase producers and high MnP producers. *Leptosphaerulina* sp. belongs to high laccase producers. The largest laccase activity exhibited by *Leptosphaerulina* sp. (650 U/L) is comparable with *T. versicolor* (577 U/L) and *C. dryhophyla* (335 U/L) (Baldrian and Šnajdr 2006). The MnP activity is also comparable with those species; however, the MnP activity in *Leptosphaerulina* sp. was small compared with other microorganisms, such as *F. trogii* (559 U/L) or *Geotrichum* sp. (300 U/L) (Máximo et al. 2003; Kariminiaae-Hamedaani et al. 2007). In turquoise blue experiment, the largest enzymatic production was observed at day 9 with a laccase activity of 512 U/L and a MnP activity of 100 U/L. The laccase activity dominance, in the treatment of phthalocyanine dyes, has been described in other fungal strains, such as*Gloeophyllum odoratum* (104 U/L), *Trametes pubescens* (391 U/L) and *F. trogii* (400 U/L) (Anastasi et al.2010; Yesilada et al. 2010). The enzymatic production was observed during all the kinetic; nevertheless, the highest enzymatic activities of laccase and MnP were observed on day 13 with 647 U/L and 90 U/L, respectively. In the degradation of monoazoic dyes is not common to see a laccase production greater than MnP. Monoazoic dyes have been decolorized by fungal strains (*T. versicolor ATCC 20869*, *T. versicolor CNPR 8107*, and *I. lacteus*) with greater MnP activities than laccase (Novotný et al. 2004; Toh et al. 2003).

The positive results obtained by *Leptosphaerulina* sp. and the differences obtained between the dyes are explained by the differences in the biochemistry of the decolorization process. The azoic dyes, Remazol black and Novacron red, are degraded principally by an enzymatic process (Michniewicz et al. 2008); whereas, phthalocyanine dyes (Turquoise Blue) are degraded by a mixture of biophysical and biochemical process. In this case, Turquoise Blue is initially absorbed by the cell wall and after that degraded by the ligninolytic enzymes (Swamy and Ramsay 1999; Kirby et al. 2000). Both types of degradation have a different physiological explanation. In the azoic dyes case, the chemical structure of azoic dyes is related with the presence of phenol groups. Phenol groups are the principal substrate for the ligninolytic enzymes; therefore, the enzymatic enzymes degrade the dye by using the structural similarity between azoic dyes and lignin (Kokol et al. 2007). On the other hand, phthalocyanine dyes are adsorbed in the membrane because they have a copper atom in their structure, copper atoms tend to be absorbed by ligninolytic fungus to further include them in the structure of their enzymes, in this case the dye molecule is absorbed and further degraded by the enzymatic enzymes (Jarosz-Wilkołazka et al. 2002).

The *Leptosphaerulina* sp. lyophilized-extracts (Fig. 4) transformed enzymatically all the dyes; however, this decolorization was dependent of the dye. *Leptosphaerulina* sp. enzymatic extracts achieved similar levels of remazol black decolorization than the commercial enzyme (80 %); however, this decolorization was less than the obtained by enzymatic extracts from *Pleurotus sajor-caju* and *Ganoderma lucidum*, using HBT as mediator (Murugesan et al. 2007). The enzymatic extracts transformed the remazol red into a reddish colour, this reddish coloration was described in the decolorization of Remazol Black by enzymatic extracts from *Pleurotus sajor-caju* and *Ganoderma lucidum* using N,N'-Bis-(1H-tetrazol-5-yl)-hydrazine (HBT) as mediator (Murugesan et al. 2007). In Turquoise Blue, the *Leptosphaerulina* sp. enzymatic extracts produced a decolorization of 56 % (Fig. 4). This enzymatic decolorization is similar to the obtained by extracts of *Dichomitus squalens* (60 %) over the copper (II) phthalocyanine dye (Svobodová et al. 2008). The low decolorization of monoazoic dyes by enzymatic extracts has been reported in acid yellow 49 and acid yellow 129 (Ciullini et al. 2008). In the future, the decolorization using*Leptosphaerulina* sp. enzymatic extracts can be improved by incrementing the MnP activity in the extracts and employing other types of chemical mediators, as HBT or vanillin.

Leptosphaerulina sp. displayed high decolorizing activity in the discharge and dyeing bath effluent (>90 %). This level of decolorization was achieved by the optimization of different factors. The principal ones were pH and effluent concentration. The importance of neutral or acid pH to realize the decolorization has been observed in *P. sajor-caju*, *B. adusta*, and *I. lacteus* (Mohorčič et al. 2006; Novotný et al. 2011). The dye concentration in the effluent was 50 times larger than the dye concentration tested in the previous experiments. *Leptosphaerulina* sp. was inhibited by the large dye concentration; however, at lower concentrations, *Leptosphaerulina* sp. decolorized efficiently the dyeing effluent. Effluent dilution reduces the growth and enzymatic production inhibitions (Mohorčič et al. 2006). Other factors such as nitrogen and glucose supplementation and initial inoculum also affected the decolorization process; however, they were less relevant than pH and effluent concentration. Glucose supplementation has been proved as an important factor in order to bio-remediate textile effluents with ligninolytic fungi (Novotný et al. 2011; Pakshirajan et al. 2011). In contrast, nitrogen supplementation had contradictory effects. *Aspergillus fumigatus* XC6 (Jin et al. 2007) and *Phanerochaete chrysosporium* (Pakshirajan et al.2011) exhibited large decolorization, when nitrogen was supplemented; whereas, *C. versicolor* showed less decolorization after nitrogen supplementation (Asgher et al. 2009).

5. CONCLUSIONS

From 92 Colombian fungal isolates, *Leptosphaerulina* sp. exhibited the best decolorization percentage (>90 %) in solid and liquid cultures, and in agitated and un-agitated conditions. Additionally,*Leptosphaerulina* sp. decolorized two real effluents from textile industries (>90 %) under conditions of low pH and glucose/nitrogen supplementation. These decolorizations were associated with the production of significant quantities of laccase (650 U/L) and MnP (100 U/L). *Leptosphaerulina* sp. enzymatic extracts decolorized (20–80 %) the three dyes; however, they needed the presence of ABTS as mediator. The cytotoxicity test and the chromatographic techniques demonstrated the dyes' biochemical transformation and the absence of negative effects in the compounds produced by this biochemical process.*Leptosphaerulina* sp. is an interesting microorganism for future biotechnological applications, specially to treat waters from the textile industry.

AUTHORS' CONTRIBUTIONS

JP was responsible of experiment's conception and design, acquisition of data, analysis and interpretation of data, and manuscript drafting. XC was responsible of microbiological and molecular experimentation. SO was responsible of microbiological and enzymatic experiments. MY was responsible of analysis and interpretation of data and manuscript drafting and approval. AM was responsible of final manuscript approval. All authors read and approved the final manuscript.

ACKNOWLEDGEMENTS

The authors acknowledge to The Research Direction of the National University of Colombia at Medellin (DIME) and the Biotechnology Postgraduate Program for the financial support to this project. Fabricato-Tejicondor S.A. for providing the dyes and effluents used in the study. The Mycotoxins, analysis and environmental processes, and molecular and cellular biology laboratories at the Faculty of Science at the National University of Colombia, at Medellin, for the technical and methodological support.

REFERENCES

1. Abadulla E, Tzanov T, Costa S, Robra KH, Cavaco-Paulo A, Gubitz GM (2000) Decolorization and detoxification of textile dyes with a laccase from *Trametes hirsuta*. Appl Environ Microbiol 66:3357–3362

2. Anastasi A, Prigione V, Varese GC (2010) Industrial dye degradation and detoxification by basidiomycetes belonging to different eco-physiological groups. J Hazard Mater 177:260–267.

3. Angel TM (2002) Molecular biology and structure-function of lignin-degrading heme peroxidases. Enzyme Microb Technol 30:425–444.

4. Archibald FS, Bourbonnais R, Jurasek L, Paice MG, Reid ID (1997) Kraft pulp bleaching and delignification by*Trametes versicolor*. J Biotechnol 53:215–236.

5. Asgher M, Azim N, Bhatti HN (2009) Decolorization of practical textile industry effluents by white rot fungus*Coriolus versicolor* IBL-04. Biochem Eng J 47:61–65

6. Axelsson J, Nilsson U, Terrazas E, Aliaga TA, Welander U (2006) Decolorization of the textile dyes Reactive Red 2 and Reactive Blue 4 using *Bjerkandera* sp. Strain BOL 13 in a continuous rotating biological contactor reactor. Enzyme Microb Technol 39:32–37

7. Babot ED, Rico A, Rencoret J, Kalum L, Lund H, Romero J, del Río JC, Martínez ÁT, Gutiérrez A (2011) Towards industrially-feasible delignification and pitch removal by treating paper pulp with *Myceliophthora thermophila* laccase and a phenolic mediator. Bioresour Technol 102:6717–6722.

8. Baldrian P, Šnajdr J (2006) Production of ligninolytic enzymes by litter-decomposing fungi and their ability to decolorize synthetic dyes. Enzyme Microb Technol 39:1023–1029.

9. Bergsten-Torralba L, Nishikawa M, Baptista D, Magalhães D, Silva Md (2009) Decolorization of different textile dyes by *Penicillium simplicissimum* and toxicity evaluation after fungal treatment. Brazilian J Microbiol. 40:808–817

10. Chanagá Vera X, Plácido Escobar J, Marín Montoya M, Pérez Yepes, del Socorro María (2012) Native fungi with industrial dye degrading

potential in the Aburrá valley, Colombia. Revista Facultad Nacional de Agronomía, Medellín. 65:6811–6821

11. Ciullini I, Tilli S, Scozzafava A, Briganti F (2008) Fungal laccase, cellobiose dehydrogenase, and chemical mediators: combined actions for the decolorization of different classes of textile dyes. Bioresour Technol 99:7003–7010.
12. Costa SM, Goncalves AR, Esposito E (2005) *Ceriporiopsis subvermispota* used in delignification of sugarcane bagasse prior to soda/anthraquinone pulping. In: Davison BH, Evans BR, Finkelstein M, McMillan JD (eds) Twenty-sixth symposium on biotechnology for fuels and chemicals. Humana Press, New York, pp 695–706
13. Couto S, Toca-Herrera J (2006) Lacasses in the textile industry. Biotechnol Mol Biol Rev 1:115–120
14. Dias A, Sampaio A, Bezerra R (2007) Environmental applications of fungal and plant systems: decolourisation of textile wastewater and related dyestuffs. In: Singh S, Tripathi R (eds) Environmental Bioremediation Technologies. Springer, Heidelberg, pp 445–463
15. dos Santos AB, Cervantes FJ, van Lier JB (2007) Review paper on current technologies for decolourisation of textile wastewaters: perspectives for anaerobic biotechnology. Bioresour Technol 98:2369–2385.
16. Erkurt EA, Ünyayar A, Kumbur H (2007) Decolorization of synthetic dyes by white rot fungi, involving laccase enzyme in the process. Process Biochem 42:1429–1435.
17. Gallego A, Torres F, Robledo S, Vélez ID, Carrillo L, Muñoz DL, Quiñones W, Fonnegra R, Roldán J, Valencia L (2006) Actividad *Leishmanicida y tripanocida* de *Acacia farnesiana, Piper arieianum, P. subpedale, Sphagnum recurvum* y *vismia baccifera* ssp. *ferruginea*. Actual Biol. 28:39–49
18. Hammel KE, Cullen D (2008) Role of fungal peroxidases in biological ligninolysis. Curr Opin Plant Biol 11:349–355.
19. Hao J, Song F, Huang F, Yang C, Zhang Z, Zheng Y, Tian X (2007) Production of laccase by a newly isolated deuteromycete

fungus *Pestalotiopsis* sp. and its decolorization of azo dye. J Ind Microbiol Biotechnol 34:233–240

20. Hernández-Luna C, Gutiérrez-Soto G, Salcedo-Martínez S (2008) Screening for decolorizing basidiomycetes in Mexico. World J Microbiol Biotechnol 24:465–473.

21. Jarosz-Wilkołazka A, Kochmańska-Rdest J, Malarczyk E, Wardas W, Leonowicz A (2002) Fungi and their ability to decolourize azo and anthraquinonic dyes. Enzyme Microb Technol 30:566–572.

22. Jin X, Liu G, Xu Z, Tao W (2007) Decolorization of a dye industry effluent by *Aspergillus fumigatus* XC6. Appl Microbiol Biotechnol 74:239–243

23. Kariminiaae-Hamedaani H, Sakurai A, Sakakibara M (2007) Decolorization of synthetic dyes by a new manganese peroxidase-producing white rot fungus. Dyes Pigm 72:157–162.

24. Khindaria A, Yamazaki I, Aust SD (1996) Stabilization of the veratryl alcohol cation radical by lignin peroxidaseâe. Biochemistry 35:6418–6424.

25. Kirby N, Marchant R, McMullan G (2000) Decolourisation of synthetic textile dyes by *Phlebia tremellosa*. FEMS Microbiol Lett 188:93–96.

26. Kokol V, Doliška A, Eichlerová I, Baldrian P, Nerud F (2007) Decolorization of textile dyes by whole cultures of*Ischnoderma resinosum* and by purified laccase and Mn-peroxidase. Enzyme Microb Technol 40:1673–1677.

27. Mattinen M, Maijala P, Nousiainen P, Smeds A, Kontro J, Sipilä J, Tamminen T, Willför S, Viikari L (2011) Oxidation of lignans and lignin model compounds by laccase in aqueous solvent systems. J Mol Catal B Enzym 72:122–129.

28. Máximo C, Costa-Ferreira M (2004) Decolourisation of reactive textile dyes by *Irpex lacteus* and lignin modifying enzymes. Process Biochem 39:1475–1479.

29. Máximo C, Amorim MTP, Costa-Ferreira M (2003) Biotransformation of industrial reactive azo dyes by*Geotrichum* sp. CCMI 1019. Enzyme Microb Technol 32:145–151.

30. Michniewicz A, Ledakowicz S, Ullrich R, Hofrichter M (2008) Kinetics of the enzymatic decolorization of textile dyes by laccase from *Cerrena unicolor*. Dyes Pigm 77:295–302.

31. Mohorčič M, Teodorovič S, Golob V, Friedrich J (2006) Fungal and enzymatic decolourisation of artificial textile dye baths. Chemosphere 63:1709–1717.

32. Murugesan K, Dhamija A, Nam I, Kim Y, Chang Y (2007) Decolourization of Reactive Black 5 by laccase: optimization by response surface methodology. Dyes Pigm 75:176–184

33. Novotný Č, Svobodová K, Kasinath A, Erbanová P (2004) Biodegradation of synthetic dyes by *Irpex lacteus*under various growth conditions. Int Biodeterior Biodegrad. 54:215–223.

34. Novotný Č, Svobodová K, Benada O, Kofroňová O, Heissenberger A, Fuchs W (2011) Potential of combined fungal and bacterial treatment for color removal in textile wastewater. Bioresour Technol 102:879–888

35. Nyanhongo GS, Gomes J, Gübitz GM, Zvauya R, Read J, Steiner W (2002) Decolorization of textile dyes by laccases from a newly isolated strain of *Trametes modesta*. Water Res 36:1449–1456.

36. Pakshirajan K, Sivasankar A, Sahoo NK (2011) Decolourization of synthetic wastewater containing azo dyes by immobilized *Phanerochaete chrysosporium* in a continuously operated RBC reactor. Appl Microbiol Biotechnol 89:1223–1232

37. Park C, Lee M, Lee B, Kim S, Chase HA, Lee J, Kim S (2007) Biodegradation and biosorption for decolorization of synthetic dyes by *Funalia trogii*. Biochem Eng J 36:59–65.

38. Risna RA, Suhirman S (2002) Ligninolytic enzyme production by Polyporaceae from Lombok Indonesia. Fungal Divers 9:123–134

39. Robinson T, Chandran B, Nigam P (2001) Studies on the production of enzymes by white-rot fungi for the decolourisation of textile dyes. Enzyme Microb Technol 29:575–579.

40. Sajben-Nagy E, Manczinger L, Škrbić B, Živančev J, Antić I, Krisch J, Vágvölgyi C (2014) Characterization of an extracellular laccase

of *Leptosphaerulina chartarum*. World J Microbiol Biotechnol 30:2449–2458

41. Salony S, Mishra S, Bisaria V (2006) Production and characterization of laccase from *Cyathus bulleri* and its use in decolourization of recalcitrant textile dyes. Appl Microbiol Biotechnol 71:646–653.

42. Sergio R (2006) Laccases: blue enzymes for green chemistry. Trends Biotechnol 24:219–226

43. Simon LT, Bishop DS, Hooper GR (1979) Ultrastructure and cytochemical localization of laccase in two strains of*Leptosphaerulina briosiana* (Pollaci) Graham and Luttrell. J Bacteriol 137:537–544

44. Svobodová K, Majcherczyk A, Novotný Č, Kües U (2008) Implication of mycelium-associated laccase from *Irpex lacteus* in the decolorization of synthetic dyes. Bioresour Technol 99:463–471.

45. Swamy J, Ramsay JA (1999) The evaluation of white rot fungi in the decoloration of textile dyes. Enzyme Microb Technol 24:130–137

46. Telke AA, Kalyani DC, Dawkar VV, Govindwar SP (2009) Influence of organic and inorganic compounds on oxidoreductive decolorization of sulfonated azo dye CI Reactive Orange 16. J Hazard Mater 172:298–309

47. Toh Y, Yen JJL, Obbard JP, Ting Y (2003) Decolourisation of azo dyes by white-rot fungi (WRF) isolated in Singapore. Enzyme Microb Technol 33:569–575.

48. Weisburger JH (2002) Comments on the history and importance of aromatic and heterocyclic amines in public health. Mutat Res 506–507:9–20.

49. Xin B, Chen G, Zheng W (2010) Bioaccumulation of Cu-complex reactive dye by growing pellets of *Penicillium oxalicum* and its mechanism. Water Res 44:3565–3572

50. Yesilada O, Yildirim SC, Birhanli E, Apohan E, Asma D, Kuru F (2010) The evaluation of pre-grown mycelial pellets in decolorization of textile dyes during repeated batch process. World J Microbiol Biotechnol 26:33–39

51. Zouari-Mechichi H, Mechichi T, Dhouib A, Sayadi S, Martínez AT, Martínez MJ (2006) Laccase purification and characterization from *Trametes trogii* isolated in Tunisia: decolorization of textile dyes by the purified enzyme. Enzyme Microb Technol 39:141–148.

Chapter 7

Disperse Dyes Based on Aminothiophenes: Their Dyeing Applications on Polyester Fabrics and Their Antimicrobial Activity

Saleh Mohammed Al-Mousawi [1,*], Morsy Ahmed El-Apasery [1,2] and Huda M. Mahmoud [3]

[1]Chemistry Department, Faculty of Science, Kuwait University, P.O. Box 5969, Safat 13060, Kuwait

[2]Dyeing, Printing and Textile Auxiliaries Department, Textile Research Division, National Research Centre, Dokki, Giza 12622, Egypt

[3]Department of Biological Sciences, Faculty of Science, Kuwait University, P.O. Box 5969, Safat 13060, Kuwait

ABSTRACT

A series of monoazo disperse dyes derived from arylazothienopyridazines were synthesized. Fastness properties of dyed polyester samples were measured. Most of the dyed fabrics tested displayed excellent washing and perspiration fastness and moderate light fastness. Finally, the biological activity of the synthesized dyes against Gram positive bacteria, Gram negative bacteria and yeast were evaluated.

KEYWORDS

biological activity; polyester fabrics; disperse dyes; thienopyridazines

1. INTRODUCTION

Aminothiophene-based azo dyes are known as disperse dyes with excellent brightness of shade. This class of dyes was established as an alternative to more expensive anthraquinone dyes [1]. Thiophene-containing azo dyes have many advantages, including a colour deepening effect as an intrinsic property of the thiophene ring and small molecular structure leading to better dyeability. The heterocyclic nature of the thiophene ring has also allowed for excellent sublimation fastness on the dyed fibers [2]. A number of researchers have studied aminothiophene derivatives as azo disperse dyes in the dyeing of synthetic fibers [3,4,5,6,7] and blended polyester/wool fibers [8], This class of compounds also showed semiconducting properties and more recently, uses in optical data storage devices [9]. The synthesis of arylazothienopyridazines and their benzo-fused derivatives have been recently extensively studied [10,11]. In spite of a large number of reports on the utility of these compounds in the dye industry, to our knowledge, their corresponding arylazothienopyridazines have never been reported as potential monoazo disperse dyes. In continuation of our interest in the synthesis of arylazothieno-pyridazines and their condensed derivatives [12], this paper reports the synthesis of some arylazo-thienopyridazines using a conventional method and their application as disperse dyes on polyester fabrics. In addition the biological activity of the synthesized dyes against *Escherichia coli* and *Pseudomonas aeruginosa*(Gram negative bacteria), *Bacillus subtilis* and *Staphylococcus aureus* (Gram positive bacteria) and*Candida albicans* (yeast) was also evaluated.

2. RESULTS AND DISCUSSION

2.1. Synthesis

Arylazopyridazinone **1** reacted with dimethylformamide dimethyl acetal (DMFDMA) in refluxing DMF for 6 hours affording dihydropyridazine-4-

carbonitrile **2** that was readily converted into the pyrido[3,4-d]pyridazine-4,5-dione disperse dye **3** in 87% yield on treatment with ammonium acetate and acetic acid at reflux for 3 h.

Compound **1** readily reacted with elemental sulphur in the presence of few drops of piperidine in dioxane as reaction medium to yield arylazoaminothienopyridazine disperse dye **4** in 89% yield. An X-ray crystal structure determination of **4** confirmed with certainly its structure [13] (*cf.* Scheme 1 and Figure 1). Typical to the established behavior of thienopyridazines, compound **4** reacted readily with *N*-phenylmaleimide in a mixture of acetic acid and dioxane to yield pyrrolo[3,4-g]phthalazine disperse dye **7** via the intermediary of the [4+2] cycloadduct **6**.

Reaction of compound **4** with DMFDMA afforded the corresponding amidine disperse dye **8**. The X-ray crystal structure determination of **8** confirmed with certainly its structure [13] (*cf.*Scheme 1 and Figure 2). Compound **8** upon heating with ammonium acetate in the presence of a few drops of acetic acid afforded the pyridopyridazine disperse dye **11** via the intermediacy of the [4+2] cycloadduct **10**. Hydrolysis of the dimethylamino moiety in the latter product afforded the final product **11**. Acylating **4** by refluxing in acetic acid resulted in the formation of the acetylamino disperse dye **9** in 80% yield.

2.2. Dyeing

Disperse dyes **3**, **4**, **7–9** and **11** were applied to polyester fabrics at 2% (dye shade), using the high temperature dyeing method (HT) at 130 °C. Yellowish-brown to brown color shades were obtained. The dyeing properties on the polyester fabrics were evaluated in terms of their fastness properties (e.g., fastnesses to washing, perspiration and light). The color of dyeing on polyester fabrics is expressed in terms of CIELAB values (Table 1), and the following CIELAB coordinates were measured: lightness (L^*); chroma (C^*); hue angle (h) from 0 to 360°; a^*, whose value represents the degree of redness (positive) and greenness (negative); and b^*, whose value represents the degree of yellowness (positive) and blueness (negative).

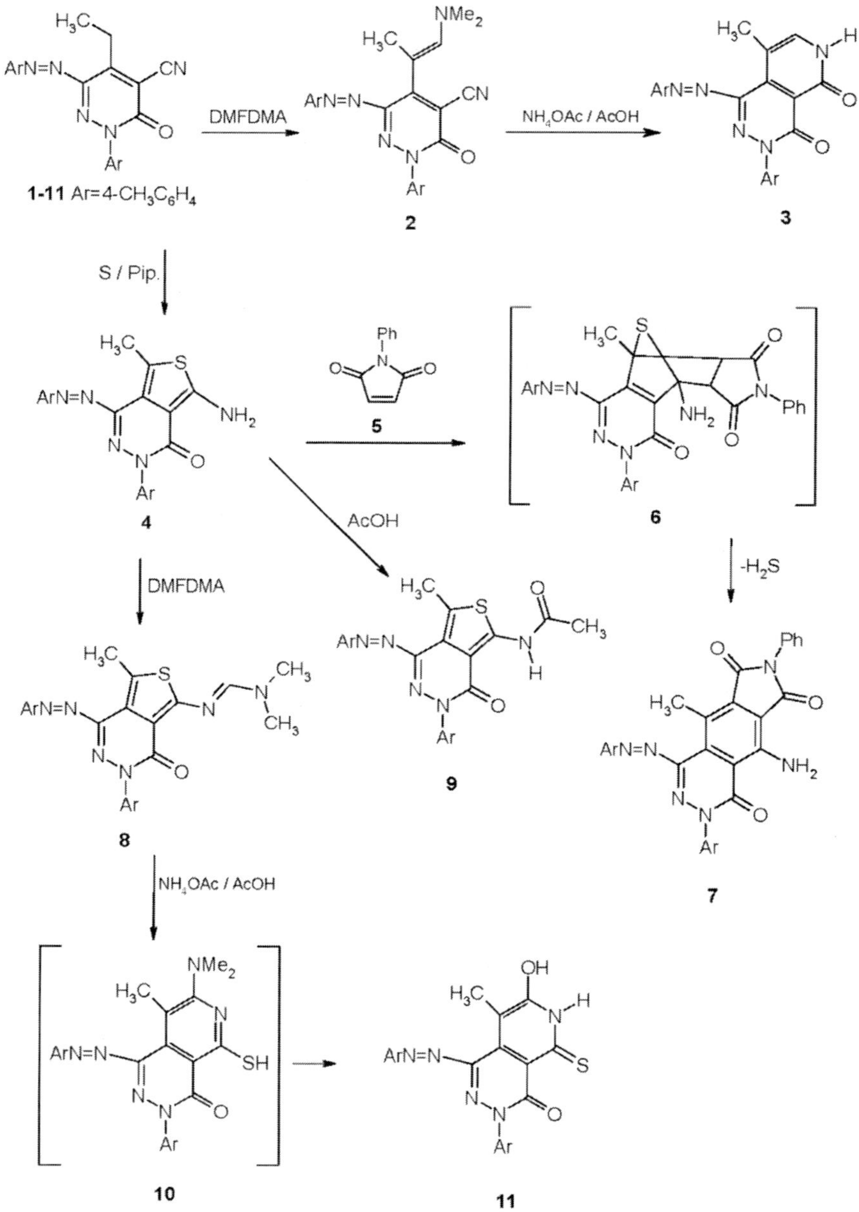

Scheme 1. Preparation of monoazo disperse dyes.

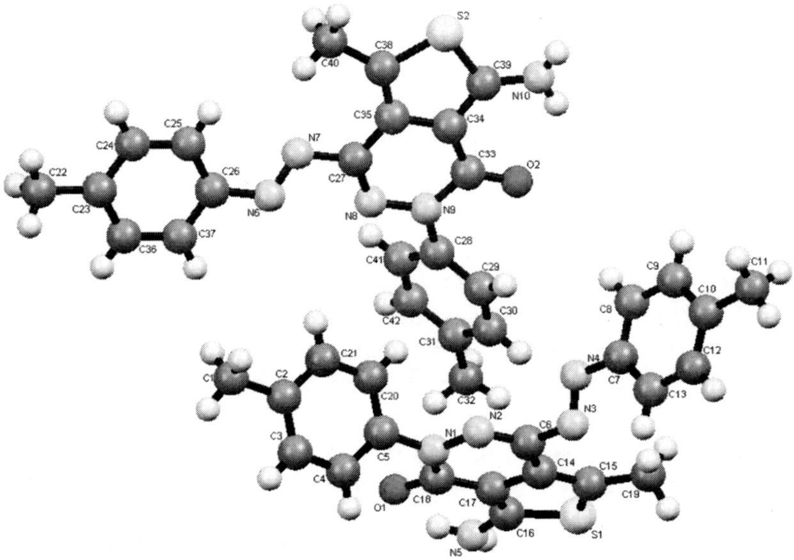

Figure 1. ORTEP plot of the X-ray crystallographic structure of**4**.

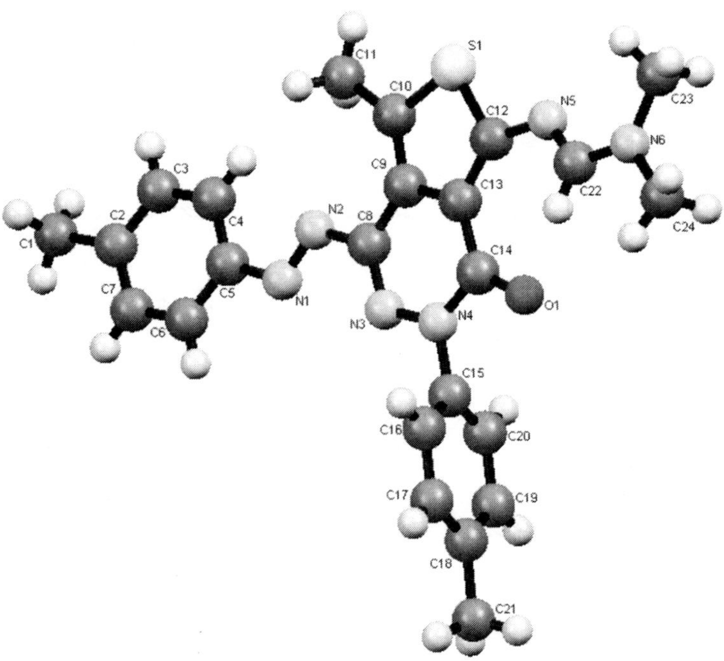

Figure 2. ORTEP plot of the X-ray crystallographic structure of**8**.

For dyeing polyester fabrics, in practical terms only disperse dyes are suitable. Through their hydrophobic properties, these dyes are capable of penetrating into the similarly hydrophobic polyester fibre. In general, the positive values of b^* (yellow–blue axis) indicated that the color hues of the all disperse dyes on the polyester fabric shifted to the yellowish directions. The results listed in Table 1 show that the dyeing fabrics have different color strengths. The difference in color strength depends on the substitutes present and/or the position of the substitutes on the structure of the synthesized dyes. A chromophore-containing compound is called a "chromogen": its color depends on the nature, number and position of the "auxochromes". Auxochromes may shift the absorption towards higher wavelengths and thus the color of the dye will deepen. Such groups are called 'bathochromes' (*i.e.*, having electron-donating substituents such as OH, NH_2 and CH_3); these are said to have a bathochromic effect. A group having the opposite, hyposochromic, effect is called a 'hypochrome' (*i.e.*, having electron-attracting substituents such as CHO, COR and NO_2). The results given in Table 1, show that the magnitude of color strength *K/S* of dye **4** (17.78) at λ_{max}464 is much higher than the color of dyes **8** and **9** (10.49 and 16.30) at λ_{max} 448 and 431, respectively. The results listed in Table 1, show also that the heteroaromatic diazo component can provide disperse dyes with bathochromic shifts due to the stabilization of the excited state.

Table 1. Shade and optical measurements of the azo disperse dyes on the polyester fabrics.

Dye No.	Color on polyester (2% shade)	*L**	*a**	*b**	*C**	*h**	*K/S*
3	Yellowish-brown	50.77	4.40	20.70	21.17	78.01	13.33
4	Brown	46.06	24.84	29.03	38.21	49.45	17.78
7	Cream	83.98	11.31	42.49	43.97	75.09	6.79
8	Light brown	68.84	22.97	39.59	45.77	59.88	10.49
9	Brownish-yellow	66.03	22.50	60.44	64.49	69.58	16.30
11	Light cream	82.96	9.55	33.54	34.87	74.11	6.76

The physical data for the dyed fabrics given in Table 2, shows that the disperse dyeing displayed moderate fastness levels to light. The light fastness of each of the dyes was measured by employing the standard method for determination of color fastness of textiles. Several reports [14] suggest that fading of azo dyes is mainly a consequence of decomposition of the –N=N– moiety either by oxidation, reduction or photolysis. The rates of these processes should be sensitive to the chemical structure of the dye, the type of substrate and treatment conditions. Since the dyed substrate employed in this study is a polyester fabrics (*i.e.*, non-proteinic), the fading process likely occurs by oxidation [15]. The ease of oxidation of azo linkages should be a function of electron density. Therefore, electron donating substituents on this moiety should increase the fading rate while electron withdrawing groups should decrease the rate. This proposal is in agreement with the observed results (Table 2) which demonstrate that the presence of a hydroxyl group in dye **11**causes decrease of light fastness to 3. On the other hand, high light fastness of dye **7** to 6, is may be attributed to hydrogen bonding between the amino group and carbonyl group cause aggregation of this dye sequent, the fading is decreased [16].

In addition the results obtained showed that dyed fabrics have excellent fastness levels to washing and perspiration fastness properties that may be due to the absence of solubilizing groups, which affects solubility, and wash ability of the dye-out of the fabrics or to the size of the dye molecule is considered relatively big.

Table 2. Fastness properties of monoazo disperse dyes on polyester fabrics.

Dye No.	Wash fastness			Perspiration fastness						Light fastness
				Alkaline			Acidic			
	Alt	SC	SW	Alt	SC	SW	Alt	SC	SW	
3	5	5	5	5	5	5	5	5	5	2
4	5	5	5	5	4–5	5	5	5	5	2
7	5	5	5	5	5	5	5	5	5	6
8	5	4–5	5	5	4–5	5	5	4–5	5	3
9	5	5	5	5	5	5	5	5	5	3
11	5	5	5	5	5	5	5	5	5	3

Alt = alteration; SC = staining on cotton; SW = staining on wool.

2.3. Antimicrobial Activities

The antimicrobial activities of the synthesized disperse dyes were screened against selected bacteria and fungi by the agar well diffusion method and their inhibition zones diameters, given inTable 3, reveal that the tested dyes **3** and **9** showed positive antimicrobial activities against at four of the tested microorganisms. Dye **3** showed strong activities (>10 mm inhibition zone) against *E. coli*, whereas dyes **7**, **8** and **11** showed no activities against any of the tested microorganisms used in the study. Only dye **9** showed a significant inhibition zone <10 mm against *Candida albicans*.

Table 3. Diameter of the zones of inhibition of the tested disperse dyes against Gram positive, Gram negative bacteria and yeast.

Dye No.	Inhibition zone diameter				
	G+ve bacteria		*G-ve* bacteria		Fungi
	B. subtilis	*S. aureus*	*E. coli*	*P. aeruginosa*	*C. albicans*
	Mean ± SD	Mean ± SD	Mean ± SD	Mean ± SD	Mean ± SD
3	8.4 ± 0.7	6.6 ± 0.2	11.2 ± 0.2	8.6 ± 1.3	-
7	-	-	-	-	-
8	-	-	-	-	-
9	5.1 ± 0.5	5.1 ± 0.2	7.4 ± 0.8	6 ± 0.3	6.4 ± 0.4
11	-	-	-	-	-
Ampicillin *	10 ± 0.5	3.5 ± 0.2	10.8 ± 0.7	9.5 ± 2.6	
Cycloheximide **					-

(-) no inhibition; * Ampicillin: Antibacterial (100 $mg{\cdot}mL^{-1}$); ** Cycloheximide: Antifungal (100 $mg{\cdot}mL^{-1}$); SD = Standard Deviation.

In addition, disperse dyes **3** and **9** (Figure 3, Figure 4) showed cytolytic effect even after six days of incubation, there were no growth recorded in the inhibited zone for all five tested microbes.

Currently, we are inspecting the biological activity of the polyester fabrics dyed with aminothiophene dyes against Gram positive bacteria, Gram negative bacteria and yeast.

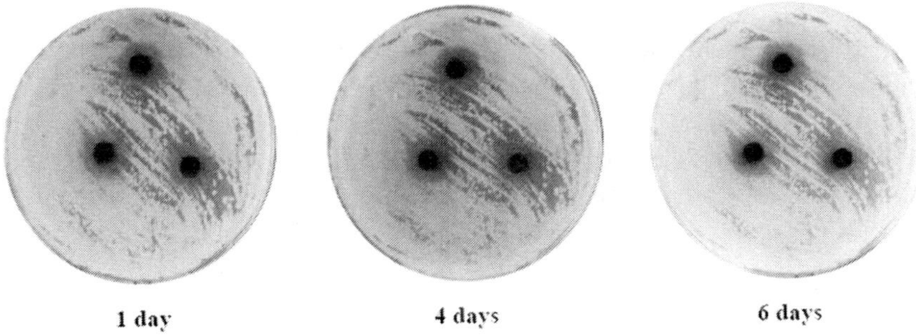

Figure 3. *E. coli*. treated with 100 mg/mL of dye **3** after 1, 4, and 6 days of incubation.

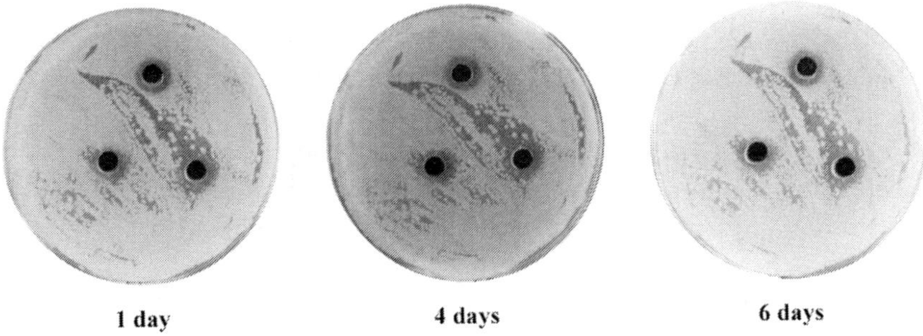

Figure 4. *E. coli*. treated with 100 mg/mL of dye **9** after 1, 4, and 6 days of incubation.

3. EXPERIMENTAL

3.1. General

Melting points were recorded on a Gallenkamp apparatus. IR spectra were recorded using KBr pellets on a JASCO FTIR-6300 FT-IR spectrophotometer. ^{1}H- and ^{13}C-NMR spectra were recorded on Bruker DPX 400 MHz or AvanceII 600 MHz super-conducting NMR spectrometers with proton spectra measured at 400, 600 MHz and carbon spectra at 100 and 150 MHz, respectively. Mass spectra were measured on a high resolution GC/MS DFS-Thermo instrument. Microanalyses were performed on an Elementar-Vario Microcube analyzer. The crystal structure of compounds **4** and **8**were determined by Bruker AXS X8 Prospector Single Crystal X-Ray Diffractometer at Kuwait University. The dyeing of polyester fabrics were conducted using LINITEST+Laboratory High Temperature Dyeing and Fastness System (ATLAS, Lichtenstein, Germany). The colorimetric parameters of the dyed polyester fabrics were determined on a reflectance spectrophotometer (UltraScan PRO D65, HunterLab, Virginia, VA, USA).

6-p-Tolyldiazenyl)-5-(1-(dimethylamino)prop-1-en-2-yl)-3-oxo-2-p-tolyl-2,3-dihydropyridazine-4 carbo-nitrile (**2**). A solution of compound **1** (3.55 g, 10 mmol) and DMFDMA (1.19 g, 10 mmol) was refluxed for 2 h then allowed to cool to room temperature. The solid product obtained was crystallized from dioxane to afford compound **2** as a yellow powder (73%), m.p. 292–294 °C; IR (KBr): 2237 (CN), 1655 (CO) cm^{-1}; ^{1}H-NMR (DMSO-d_6): δ = 2.08 (s, 3H, CH_3), 2.23 (s, 6H, 2(*p*-tolyl-CH_3)), 2.27 (s, 6H, N-$(CH_3)_2$), 6.80 (d, 2H, *J* = 7.7 Hz, *p*-tolyl-H), 7.01 (d, 2H, *J* = 6.6 Hz, *p*-tolyl-H), 7.07–7.13 (m, 2H, *p*-tolyl-H), 7.21 (d, 2H, *J* = 8.1 Hz, *p*-tolyl-H), 7.32 (d, 1H, *J* = 7.9 Hz, vinylic-H). MS, *m/z* (%), 412 ($[M]^+$, 45), UV/Vis at λ_{max}(DMF) = 432 nm. Anal. Calcd. For $C_{24}H_{24}N_6O$: C, 69.88; H, 5.86; N, 20.37. Found C, 69.91; H, 5.44; N, 19.73.

8-Methyl-3-p-tolyl-1-(p-tolyldiazenyl)pyrido[3,4-d]pyridazine-4,5(3H,6H)-dione(**3**). A solution of compound **2** (4.32 g, 10 mmol), acetic acid (15 mL) and ammonium acetate (1.15 g, 15 mmol) was refluxed for 3 h, then allowed to cool to room temperature. The solid product so formed was collected by filtration and crystallized from acetic acid. Compound **3** was obtained as a light brown powder (87%), m.p. 226–228 °C; IR (KBr): 3451 (NH), 1706, 1659 (CO) cm^{-1}; ^{1}H-NMR (DMSO-d_6): δ = 2.16 (s, 3H, CH_3), 2.31 (s, 6H, 2(*p*-tolyl-CH_3)), 7.10–7.40 (m, 4H, *p*-tolyl-H), 7.46 (d, 2H, *J* = 8.4 Hz,*p*-tolyl-H), 7.52 (d, 2H, *J* = 7.2 Hz, *p*-tolyl-H), 8.79 (s, 1H, pyridyl-H). 9.91 (s, 1H, NH).MS, *m/z* (%), 383 ($[M-2]^+$, 32), UV/Vis at λ_{max}(DMF) = 590. Anal. Calcd. For $C_{22}H_{19}N_5O_2$: C, 68.56; H, 4.97; N, 18.17. Found C, 68.58; H, 4.99; N, 18.26.

7-Amino-5-methyl-2-p-tolyl-4-p-tolylazo-2H-thieno[3,4-d]pyridazin-1-one (**4**). A mixture of compound**1** (3.55 g, 10 mmol) and elemental sulphur (0.32 g, 10 mmol) and a few drops of piperidine in dioxane (10 mL) was refluxed for 4 h then allowed to cool to room temperature. The crude product was poured onto water, the solid product, so formed, was collected by filtration and crystallized from ethanol. Compound **4** was obtained as wine red crystals (89%), m.p. 209 °C (Lit. [12] m.p. 208–209 °C), IR (KBr): 3321, 3278 (NH_2), 1642 (CO) cm^{-1}; ^{1}H-NMR (DMSO-d_6): δ = 2.30 (s, 3H, CH_3), 2.37 (s, 3H, *p*-tolyl-CH_3), 2.40 (s, 3H, *p*-tolyl-CH_3), 7.21 (d, 2H, *J* = 10.2 Hz, *p*-tolyl-H), 7.32 (d, 2H, *J* = 9.6 Hz, *p*-tolyl-H), 7.38 (s, 2H, NH_2), 7.39

(d, 2H, J = 7.2 Hz, p-tolyl-H), 7.78 (d, 2H, J = 7.2 Hz, p-tolyl-H). ^{13}C NMR (DMSO-d_6): δ = 160.77 (CO), 159.2, 150.4, 150.1, 144.0, 139.5, 136.9, 130.7, 129.5, 126.3, 123.6, 121.9. 116.7, 104.7, 21.5 (p-tolyl-CH_3), 21.0 (p-tolyl-CH_3), 14.7 (CH_3). MS (EI) m/z = 389 $(M]^+$, 100); UV/Vis at λ_{max}(DMF) = 464 nm. Anal. Calcd for $C_{21}H_{19}N_5OS$: C, 64.76; H, 4.92; N, 17.98; S, 8.23. Found: C, 65.09; H, 5.08; N, 17.90; S, 8.21.

9-Amino-5-methyl-7-phenyl-2-p-tolyl-4-p-tolylazo-2H-pyrrolo[3,4-g]phthalazine-1,6,8-trione (**7**). A solution of compound **4** (3.89 g, 10 mmol), N-phenylmaleimide (1.73 g, 10 mmol) and a few drops of acetic acid was refluxed in dioxane (15 mL) for 10 h. The solid products, so formed, were collected by filtration and crystallized from DMF. Compound **7** was obtained as orange crystals (79%), m.p. 302–304 °C) (Lit. [12] m.p. 300–302 °C), IR (KBr): 3435, 3310 (NH_2),1752, 1704, 1645 (CO) cm^{-1}; ^{1}H-NMR (DMSO-d_6): δ = 2.42 (s, 3H, CH_3), 2.48 (s, 3H, p-tolyl-CH_3), 2.75 (s, 3H, p-tolyl-CH_3), 7.36 (d, 2H, J = 7.8 Hz, p-tolyl-H), 7.50 (d, 4H, J = 8.1 Hz, p-tolyl-H), 7.57–7.64 (m, 5H, phenyl-H), 7.94 (d, 2H, J = 7.7 Hz, p-tolyl-H), 9.47 (s, 2H, NH_2). MS (EI) m/z = 528 ($[M]^+$, UV/Vis atλ_{max}(DMF) = 471 nm. Anal. Calcd for $C_{31}H_{24}N_6 6O_3$: C, 70.44; H, 4.58; N, 15.90. Found: C, 69.99; H, 4.58; N, 15.96.

N,N-Dimethyl-N'-(7-methyl-4-oxo-3-p-tolyl-1-p-tolylazo-3,4-dihydrothieno[3,4-d]pyridazin-5-yl)-formamidine (**8**). A mixture of compound **4** (3.89 g, 10 mmol) and DMFDMA (1.19 g, 10 mmol) in dimethylformamide (15 mL), was refluxed for 1 h. The solvent was evaporated and then the residue was washed with ethanol. The solid product, so formed, was collected by filtration and crystallized from dimethylformamide. Compound **8** was obtained as wine red crystals (93%), m.p. 307 °C (Lit. [12] m.p. 203–205 °C), IR (KBr): 1651 (CO) cm^{-1}. ^{1}H-NMR (DMSO-d_6): δ = 2.31 (s, 3H, CH_3), 2.39 (s, 3H, p-tolyl-CH_3), 2.53 (s, 3H, p-tolyl-CH_3), 2.96 (s, 3H, N-CH_3), 3.04 (s, 3H, N-CH_3), 7.22 (d, 2H, J = 9.0 Hz, p-tolyl-H), 7.30 (d, 2H, J = 9.0 Hz, p-tolyl-H), 7.41 (d, 2H, J = 7.8 Hz, p-tolyl-H), 7.80 (d, 2H, J = 8.4 Hz, p-tolyl-H), 7.92 (s, 1H, amidine-H). ^{13}C NMR (DMSO-d_6): δ = 163.7 (CO), 157.5, 157.3, 150.5, 149.9, 144.0, 139.2, 137.1, 130.7, 129.3, 126.8, 125.4, 123.8, 123.7, 112.9, 34.7 (N-CH_3), 21.5 (p-tolyl-CH_3), 21.0 (p-tolyl-CH_3), 15.8 (CH_3). MS (EI) m/z = 444 $(M]^+$, 100), UV/Vis atλ_{max}(DMF) = 448 nm. Anal. Calcd for $C_{24}H_{24}N_6OS$: C, 64.84; H, 5.44; N, 18.90; S, 7.21. Found: C, 64.58; H, 5.59; N, 18.70; S, 7.16.

N-(7-Methyl-4-oxo-3-p-tolyl-1-p-tolylazo-3,4-dihydrothieno[3,4-d]pyridazin-5-yl)acetamide (**9**). A solution of compound **4** (3.89 g, 10 mmol) in acetic acid (20 mL) was refluxed for 3h. The solvent was evaporated and then the residue was washed with ethanol. The solid product, so formed, was collected by filtration and crystallized from dioxane. Compound **9** was obtained as brown crystals (80%), m.p. 209 °C) (Lit. [12] m.p. 207–209 °C), IR (KBr): 3270 (NH), 1673, 1641 (CO) cm^{-1}. ^1H-NMR (DMSO-d_6): δ = 2.30 (s, 3H,$COCH_3$), 2.26 (s, 3H, *p*-tolyl-CH_3), 2.33 (s, 3H, *p*-tolyl-CH_3), 2.39 (s, 3H, CH_3), 7.27 (d, 2H, *J* = 9.0 Hz, *p*-tolyl-H), 7.40–7.44 (m, 4H, *p*-tolyl-H), 7.83 (d, 2H, *J* = 7.8 Hz, *p*-tolyl-H), 11.03 (s, 1H, NH). ^{13}C NMR (DMSO-d_6): δ = 167.6 (CO), 158.3, 153.5, 149.9, 144.4, 142.7, 138.3, 137.4, 130.8, 129.5, 126.6, 123.8, 120.8, 119.8, 112.4, 23.1 (*p*-tolyl-CH_3), 21.1 (*p*-tolyl-CH_3), 16.5 (CH_3), 14.8 (CH_3). MS (EI) *m/z* = 431 $(M]^+$, 100), UV/Vis at λ_{max}(DMF) = 431 nm. Anal. Calcd for $C_{23}H_{21}N_5O_2S$: C, 64.02; H, 4.91; N, 16.23; S, 7.43. Found: C, 63.88; H, 4.89; N, 16.36; S, 7.34.

7-Hydroxy-8-methyl-5-thioxo-3-p-tolyl-1-p-tolylazo-5,6-dihydro-3H-pyrido [3,4-d]pyridazin-4-one (**11**). A mixture of compound **8** (4.44 g, 10 mmol) acetic acid (10 mL) and ammonium acetate (3.45 g, 30 mmol) was refluxed for 3 h. The crude product was poured onto water, the solid product, so formed, was collected by filtration and crystallized from dioxane/ethanol (3:1). Compound **11** was obtained as a light violet powder (68%), m.p. 297 °C (Lit. [12] m.p. 295–297 °C), IR (KBr): 3440 (OH), 3275 (NH), 1681, 1641 (CO) cm^{-1}. ^1H-NMR (DMSO-d_6): δ = 2.37 (s, 3H, CH_3), 2.43 (s, 3H, *p*-tolyl-CH_3), 2.66 (s, 3H, *p*-tolyl-CH_3), 7.29 (d, 2H, *J* = 7.8 Hz, *p*-tolyl-H), 7.44 (t, 4H, *J* = 7.4 Hz, *p*-tolyl-H), 7.87 (d, 2H, *J* = 7.8 Hz, *p*-tolyl-H), 8.61 (s, 1H, NH), 11.7 (s, 1H, OH). ^{13}C NMR (DMSO-d_6): δ = 157.8, 151.3, 151.0, 144.4, 141.1, 138.4, 138.2, 130.6, 130.2, 130.0, 129.6, 126.4, 125.6, 124.5, 120.2, 22.3 (*p*-tolyl-CH_3), 21.8 (*p*-tolyl-CH_3), 15.7 (CH_3). MS (EI) *m/z* = 417 $(M]^+$, 100), Anal. Calcd for $C_{22}H1_{19}N_5O_2S$: C, 63.29; H, 4.59; N, 16.78; S, 7.68. Found: C, 63.26; H, 4.78; N, 16.75; S, 7.56.

3.2. High Temperature Dyeing Method (HT)

3.2.1. Materials

Polyester 100% (150 130 g/m^2, 70/2 denier) was used. The fabric was treated before dyeing with a solution containing non-ionic detergent (Sera Wash M-RK, 5 g/L) and sodium carbonate (2 g/L) in a ratio of 50:1 at 60 °C for 30 min, then thoroughly washed with water and air dried at room temperature.

3.2.2. Dyeing

The dye baths were prepared from the dye (2% weight of fibre) to a final liquor of 50:1, w/w. The pH value of the bath was adjusted to 4.5–5 with acetic acid (10%) in the presence of a 1:1 ratio of the dispersing agent (Sera Gal P-LP). The temperature was raised to 130 °C at the rate of 7 °C/min, and dyeing continued for 60 min. After dyeing, the fabrics were thoroughly washed and then subjected to a surface reduction clearing [(2 g NaOH + 2 g sodium hydrosulphite)/L]. The samples were heated in this solution for 30 min at 80 °C and then thoroughly washed and air-dried.

3.3. Color Measurements and Analyses

3.3.1. Color Measurements

The colorimetric parameters (Table 1) of the dyed polyester fabrics were determined on a reflectance spectrophotometer. The color yields of the dyed samples were determined by using the light reflectance technique performed on (UltraScan PRO D65) UV/VIS Spectrophotometer. The color strengths, expressed as *K/S* values, were determined by applying the Kubelka-Mink equation as follows:

$$K/S = [(1 - R)^2/2R] - [(1 - R_o)^2/2R_o]$$

where R = decimal fraction of the reflectance of the dyed fabric; R_o = decimal fraction of the reflectance of the undyed fabric; K = absorption coefficient; S = scattering coefficient.

3.3.2. Fastness Testing

3.3.2.1. Fastness to Washing

After washing using 2 g/L of the nonionic detergent Hostapal CV at 80 °C for 15 min, the dyed fabrics were tested by using standard methods [17]. A specimen of dyed polyester fabric was stitched between two pieces of undyed cotton and wool fabrics, all of equal length, and then washed at 95 °C for 30 min. The staining on the undyed adjacent fabrics was assessed according to the following gray scale: 1—poor, 2—fair, 3—moderate, 4—good, 5—excellent.

3.3.2.2. Fastness to Perspiration

The samples were prepared by stitching a piece of dyed polyester fabric between two pieces of cotton and wool fabrics, all of equal length, and then immersed in the acid or alkaline solution for 30 min. The staining on the undyed adjacent fabrics was assessed according to the following gray scale: 1—poor, 2—fair, 3—moderate, 4—good, 5—excellent. The acid solution (pH = 4.5) contains sodium chloride (10 g/L), sodium dihydrogen orthophosphate (1 g/L) and histidine monohydrochloride (0.25 g/L). The alkaline solution (pH = 9.5) contains sodium chloride (10 g/L), disodium orthophosphate (1 g/L) and histidine monohydrochloride (0.25 g/L).

3.3.2.3. Fastness to Light

Light fastness was determined by exposing the dyed polyester fabrics on a Xenotest 150 (Original Hanau, Lichtenstein, Germany). Chamber temperature: 25–30 °C, black panel temperature: 60 °C, relative humidity: 50%–60%, dark glass UV filter system) for 40 h. The changes in color were assessed according to the following blue scale: 1—poor, 3—moderate, 4—good, 6—very good, 8—excellent.

3.4. Antimicrobial Activities Test

The antimicrobial activities of different arylazothienopyridazines disperse dyes were tested using Agar-well diffusion technique [18] against five different microbial cultures. Pure cultures of*Escherichia coli* and *pseudomonas aeruginosa* (Gram negative bacteria), *Bacillus subtilis* and*Staphylococcus aureus* (Gram positive bacteria) and *Candida albicans* (yeast) were used in the test. An aliquot of 0.1 mL of each bacterial strain was inoculated and spread on nutrient agar (NA) while 0.1 mL of the yeast was spread on potato dextrose agar (PDA). The inoculated plates were supplied with 100 μL of each of the tested dyes with a total final concentration of 10 mg mL^{-1}. The dyes were included in 4 mm wells produced by sterile cork borer. The NA plates were incubated at 37 °C for 24 h while PDA plates were incubated at 25 °C for 24–48 h. The zones of inhibition around the wells were determined and the average based on three replica was recorded. Cycloheximide and ampicillin were both used as references in the experiment, where cycloheximide known to inhibit eukaryotic organisms while ampicillin inhibits prokaryote organisms. All plates were kept for six days after inoculation and the changes in the inhibition zone was monitored and documented by photography in order to determine on the cytolytic and cytostatic effect of the tested disperse dyes.

4. CONCLUSIONS

In summary, a series of monoazo disperse dyes based on arylazothienopyridazines were synthesized. The dyes produced in this manner were then applied to polyester fabrics by using a high temperature dyeing method at 130 °C. The dyed polyester fabrics, which display yellowish-brown to brown hues, displayed excellent washing and perspiration fastness and moderate light fastness. Finally, the biological activity of the synthesized compounds against Gram positive bacteria, Gram negative bacteria and yeast was tested.

ACKNOWLEDGMENTS

The authors are grateful to M.H. Elnagdi for his help and valuable advices. Support of this work provided by Kuwait University through a research grant (SC 05/09) and the facilities of GF-S, supported by research grants (GS03/08), (GS01/01), (GS01/03) and (GS01/05) are highly appreciated.

REFERENCES AND NOTES

1. Towns, A.D. Developments in azo disperse dyes derived from heterocyclic diazo components. *Dyes Pigm.* **1999**, *42*, 3–28.

2. Hallas, G.; Towns, A.D. Dyes derived from aminothiophenes. Part 6: Application of some nitro-substituted thiophene-based azo disperse dyes to hydrophobic fibres. *Dyes Pigm.* **1997**,*35*, 45–55.

3. Maradiya, H.R.; Patel, V.S. Thiophene based monoazo disperse dyes for polyester fabric. *J. Serb. Chem. Soc.* **2002**, *67*, 17–25.

4. Bhatti, H.S.; Seshadri, S. Novel disperse dyes from benzo[b]thiophene-3(2*H*)-one-1,1-dioxide and ethylbenzo[b]thien-3(2*H*)-ylidenecy-anoacetate, S,S-dioxide:synthesis and properties.*Dyes Pigm.* **2004**, *62*, 83–92.

5. Naik, S.J.; Halkar, U.P. Synthesis of 4,5,6,7-tetrahydrobenzo [b] thiophene based azo disperse dyes. *Ind. J. Het. Chem.* **2006**, *15*, 213–216.

6. Rangnekar, D.W.; Telange, V.R. Synthesis of novel thieno[2,3-b]thiophene based bisazo disperse dyes. *J. Ind. Coun. Chem.* **2000**, *17*, 11–14.

7. Maradiya, H.R.; Patel, V.S. Synthesis of 2-azobenzo[b]thiophene derivatives and their application on nylon fibre. *Chem. Heterocycl. Compd.* **2002**, *38*, 281–286.

8. Afifi, T.H. One-bath dyeing of polyester/wool blend using azo dyes derived from 2-aminothiophenes. *Advan. Col. Sci. Tech.* **2003**, *6*, 63–72.

9. Abd-El-Aziz, A.S.; Afifi, T.H. Novel azo disperse dyes derived from aminothiophenes: Synthesis and UV-visible studies. *Dyes Pigm.* **2006**, *70*, 8–17.

10. Al-Mousawi, S.; Elassar, A-Z.; El-Apasery, M.A. A microwave assisted diazo coupling reaction: the synthesis of alkyl azines and thienopyridazines. *Phosphorus, Sul. Sil. Rel. Elem.***2006**, *181*, 1755–1771.

11. Al-Mousawi, S.M.; El-Apasery, M.A.; Al-Kandery, N.; Elnagdi, M.H. Synthesis of 3-arylazo-6-oxopyridazine-5-carbonitriles: A versatile precursor for condensed arylazopyridazin-6-ones. *J. Het. Chem.* **2008**, *45*, 359–364.

12. Al-Mousawi, S.M.; EL-Apasery, M.A.; Elnagdi, M.H. Studies with condensed aminothiophenes: microwave assisted cycloaddition reactions of thieno[3,4-d]pyridazinone and thieno[3,4-c]quinolinone. *Heterocycl.* **2008**, *75*, 1151–1161.

13. Crystallographic data for the structures of compounds **4** and **8** reported in this paper have been deposited with the Cambridge Crystallographic Data Centre as supplementary publications Nos. CCDC 941638 and 941639. These data can be obtained free of charge from the Cambridge Crystallographic Data Centre. Available online:

14. Al-Etaibi, A.M.; El-Apasery, M.A.; Ibrahim, M.R.; Al-Awadi, N.A. A facile synthesis of new monoazo disperse dyes derived from 4-hydroxyphenylazopyrazole-5-amines: evaluation of microwave assisted dyeing behavior. *Molecules* **2012**, *17*, 13891–13909.

15. Chipalkatti, H.R.; Desai, N.F.; Giles, C.H.; Macaulay, N. The influence of the substrate upon the light fading of azo dyes. *J. Soc. Dyers Colour* **1954**, *70*, 487–501.

16. Ritter, J.; Borst, H.U.; Lindner, T.; Hauser, M.; Brosig, S.; Bredereck, K.; Steiner, U.E.; Kuehn, D.; Kelemen, J.; Kramer, H.E.A. Substituent effects on triplet yields in aminoanthraquinones: Radiationless deactivation via intermolecular and intramolecular hydrogen bonding. *J. Photochem. Photobiol. A Chem.* **1988**, *41*, 227–244.

17. Chrysler, L.P. *Methods of Test for Color Fastness of Textiles and Leather*, 7th ed.; Bradford: London, UK, 1990.

18. Isaacson, D.M.; Kirschbaum, J. Assays of antimicrobial substances. In *Manual of Industrial Microbiology and Biotechnology*; Demain, A.L., Solomon, N.A., Eds.; ASM: Washington, DC, USA, 1986.

Chapter 8

Cyclodextrins in Textile Finishing

Bojana Voncina[1] and Vera Vivod[1]

[1] Faculty of Mechanical Engineering, University of Maribor, Maribor, Slovenia

1. INTRODUCTION

Chemical finishing is crucial for giving textiles new functionalities and making them appropriate for special applications, such as antimicrobial resistance, flame retardancy and others. Textile finishing is also an important process as it improves appearance, performance or hand.

Cyclodextrins can act as hosts and form inclusion compounds with various small molecules. Such complexes can be formed in solutions, in a solid state as well as when cyclodextrins are linked to various surfaces where they can act as permanent or temporary hosts for small molecules that provide certain desirable attributes. This characteristic makes cyclodextrin a promising reagent in textile finishing.

2. CYCLODEXTRINS

Cyclodextrins(CDs) are relevant molecules used in different applications and industries from pharmacology, cosmetics, textiles, filtration to pesticide formulations. They comprise a family of three well-known industrially produced substances. The practically important, industrially produced CDs are the α-, β-, and γ-CDs. There are some seldom used cyclic oligosaccharides as well but because of their cost are not applicable to industrial applications [1]. CDs are cyclic oligomers of a-D-glucopyranose that can be produced with the transformation of starch by certain bacterias such as Bacillus macerans [2, 3].

The preparation process of CDs consists of four principal phases: (i) culturing of the microorganism that produces the cyclodextrin glucosyl transferase enzyme (CGT-ase); (ii) separation, concentration and purification of the enzyme from the fermentation medium; (iii) enzymatical conversion of prehydrolyzed starch in mixture of cyclic and acyclic dextrins; and (iv) separation of CDs from the mixture, their purification and crystallization. CGT-ase enzymes degrade the starch and starts intramolecular reactions without the water participation. In the process, cyclic (CDs) and acyclic dextrins are originated, which are oligosaccharides of intermediate size. CDs are formed by the link between units of glucopyranose. The union is made through glycosidic oxygen bridges by α-(1,4) bonds. The purification of α- and γ-CDs increases the cost of production considerably, so that 97% of the CDs used in the market are β-CDs [2].

CDs ring structures act as hosts and form inclusion compounds with various small molecules. Such complexes can be formed in solutions, in a solid state as well as when cyclodextrins are linked to various surfaces. In all forms they can act as permanent or temporary hosts to small molecules that provide certain desirable attributes.

In the textile field CDs may have many applications such as: absorption of unpleasant odours; they can complex and release fragrances, "skin-care-active" and bioactive substances. Further, various textile materials treated with cyclodextrins could be used as selective filters for adsorption of small pollutants from waste water [4].

After the discovery of CDs scientists considered them poisonous substances and their ability for complexes formation was only considered a scientific curiosity. Later on, research on CDs proved that they are not only non-toxic but they can be helpful for protecting flavours, vitamins and natural colours [2]. CDs already play a significant role in the textile industry and can be used in dyeing, surface modification, encapsulation, washing, and preparation of polymers and in fibre spinning.

Since year 2000, β-CD has been introduced as a food additive in Germany. With respect to OECD experiments, this compound has shown no allergic impact. In the USA α-, β- and γ-CDs have obtained the GRAS list (FDA list of food additives that are 'generally recognized as safe') status and can be commercialized as such. In Japan α-, β- and γ-CDs are recognized as natural products and their commercialization in the food sector is restricted only by considerations of purity. In Australia and New Zealand α- and γ-CDs are classified as Novel Foods from 2003 and 2004, respectively. The recommendation of Joint FAO/WHO Expert Committee on Food Additives (JECFA) for a maximum level of β-CDs in foods is 5 mg/kg per day. For α- and γ-CDs no Acceptable Daily Intake (ADI) was defined because of their favourable toxicological profiles [2, 5].

Natural cyclodextrins and their hydrophilic derivatives are only able to permeate lipophilic biological membranes, such as the eye cornea, with considerable difficulty. All toxicity studies have demonstrated that orally administered cyclodextrins are practically non-toxic, due to lack of absorption from the gastrointestinal tract. The main properties of β-CD, the most important cyclodextrin in textile application are: less irritating than α-CD after i.m. injection, binds cholesterol, small amount (1-2%) is adsorbed in the upper intestinal tract, no metabolism in the upper intestinal tract, metabolised by bacteria in caecum and colon, LD50 oral rat > 5000 mg/kg, LD50 i.v., rat: between 450-790 mg/kg, however, application of high doses may be harmful and is not recommended [6, 7].

2.1. Structure Of Cyclodextrins

The three mayor CDs are crystalline, homogeneous, and non-hygroscopic substances, which are torus-shaped oligosaccharides [1, 8-10], composed in the more common forms of six to eight (α-1,4)-linked α-D-glucopyranose units (α-, β- and γ-cyclodextrin), Figure 1 schematically present the main three cyclodextrins [11]. They are of circular and conical conformation, where the height is about 800 pm. The inner diameter of the cavity varies from 500 to 800 pm.

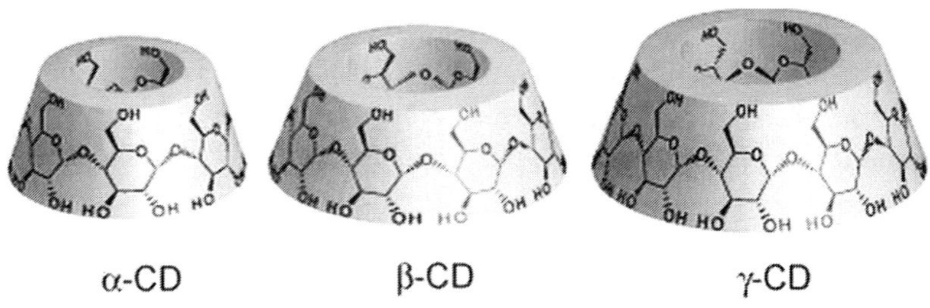

Figure 1. Schematic presentation of the main three CDs [**12**].

All glucopyranose units in the torus-like ring possess the thermodynamically favoured chair conformation because all substituents are in the equatorial position. As a consequence of the 4C1 conformation of the glucopyranose units, all secondary hydroxyl groups are situated on the lager side of the ring, whereas all the primary ones are placed on the narrower side of the ring. Hydroxyl groups on the outside of the CDs ensure good water solubility. The cavity is lined with hydrogen atoms of C3, by the glycosidic oxygen bridges and hydrogen atoms of C5. The nonbonding electron pairs of the glycosidic oxygen bridges are directed toward the inside of the cavity producing a high electron density and because of this the inner side of the cavity has some Lewis base characteristics. The C-2-OH group of one glucopyranoside unit can form a hydrogen bond with the C-3-OH group of the adjacent glucopyranose unit. In the CD molecule, a complete secondary belt is formed by these hydrogen bonds, therefore the β-CD has a rather rigid structure. Because of this arrangement, the interior of the toroids is not hydrophobic but considerably less hydrophilic than the aqueous environment and thus able to host other hydrophobic molecules. CDs behave more or less like rigid compounds with two degrees of freedom, rotation at the glucosidic links C4-O4 and C1-O4 and rotations at the O6 primary hydroxyl groups at the C5-C6 band. The intramolecular hydrogen bond formation is probably the explanation for the observation that β-CD has the lowest water solubility of all CDs. The hydrogen-bond belt is incomplete in the α-CD molecule, because one glucopyranose unit is in a distorted position. Consequently, instead of the six possible H-bonds, only four can be established fully. γ-CD is a noncoplanar with more flexible structure; therefore, it is the most soluble of the three CDs. Figure 2shows a sketch of the characteristic structural features of CDs. On the side where the secondary hydroxyl groups are situated, the diameter of the cavity is larger than on the side with the primary hydroxyls, since free rotation of the primary hydroxyls reduces the effective diameter of the cavity [13,14].

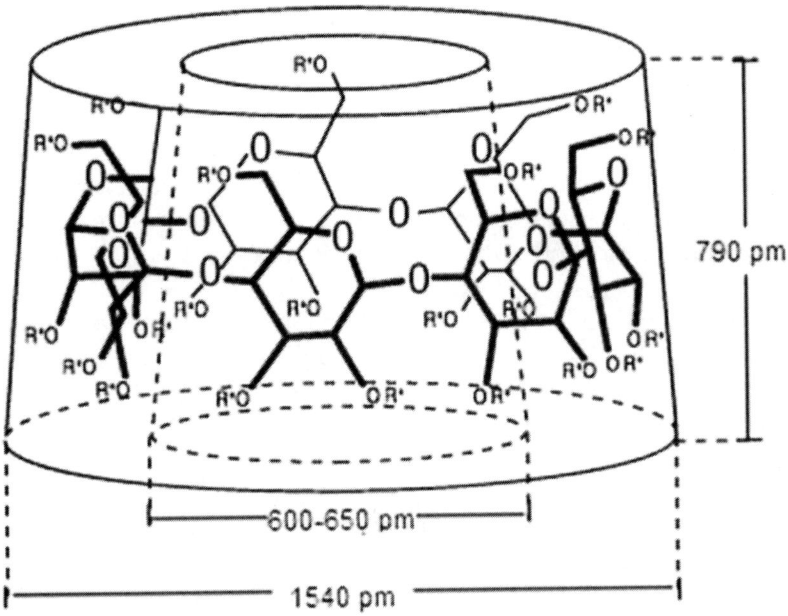

Figure 2. Schematic presentations of characteristic structural features of CDs.

By far, β-cyclodextrin (β-CD), with 7 sugar units, has been commercially the most attractive (more than 95% consumed) due to its simple synthesis, availability and price.

A β-CD molecule has a molecular weight of 1135 and a height of 750–800 pm. The internal diameter of the molecule's hole is between 600 and 680 pm, and the external diameter is 1530 pm[1, 15]. The volume of the hole is 260–265 Å^3, and the dissolution is 1.85 g/100 mL of water. The cavity is hydrophobic; the external section is hydrophilic in nature. β-CD is stable in alkali solutions and it is sensitive to acid hydrolysis [16].

2.2. Properties Of Cyclodextrins

The most notable feature of CDs is their ability to form solid inclusion complexes ("host–guest" complexes) with a very wide range of solid, liquid and gaseous compounds by a molecular complexation. The phenomenon of CD inclusion compound formation is a complicated process involving many factors playing an important role.

Complex formation is a dimensional fit between host cavity and guest molecule. The lipophilic cavity of CD molecules provides a microenvironment into which appropriately sized non-polar moieties can enter to form inclusion complexes. No covalent bonds are broken or formed during formation of the inclusion complex.

According to some authors [11, 17] hydrophobic interactions are the main driving forces for CD-based host-guest compounds. Other requirements such as steric hindrance and relation between sizes of host and guest cavities are also important. This is illustrated by the fact that not only hydrophobic interaction will lead to an association between a guest molecule and a CD but ionic solutes, such as non-associated inorganic salts, can also be involved in these complexes.

Some researchers [7] claim that the main driving force of complex formation is the release of enthalpy-rich water molecules from the cavity. The water molecules located inside the cavity cannot satisfy their hydrogen bonding potentials and therefore are of higher enthalpy. The energy of the system is lowered when these enthalpy–rich water molecules are replaced with suitable guest molecules which are less polar than water. In an aqueous solution, the slightly apolar CD cavity is occupied by water molecules which are energetically unfavoured, and therefore can be readily substituted by appropriate "guest molecules" which are less polar than water. This apolar–apolar association decreases the CD ring strain resulting in a more stable lower energy state. The dissolved CD is the "host" molecule, and the "driving force" of the complex formation is the substitution of the high-enthalpy water molecules by an appropriate "guest" molecule.

The binding of guest molecules within the host CD is not fixed or permanent but rather is a dynamic equilibrium. Binding strength depends on how well the 'host–guest' complex fits together and on specific local interactions between surface atoms. Complexes can be formed either in solution or in the crystalline state and water is typically the solvent of choice. Inclusion complexation can be accomplished in a co-solvent system and in the presence of any non-aqueous solvent [7]. Generally, one guest molecule is included in one CD molecule, although in the case of some low molecular weight molecules, more than one guest molecule may fit into the cavity, and in the case of some high molecular weight molecules, more than one CD molecules may bind to the guest. In principle, only a portion of the molecule must fit into the cavity to form a complex. CD inclusion is a stoicmetric molecular phenomenon in which usually only one guest molecule interacts with the cavity of the CD molecules to become entrapped. 1:1 complex is the simplest and most frequent case. However, 2:1, 1:2, 2:2, or even more complicated associations, and higher order equilibrium exist almost always simultaneously.

Inclusion in CDs has a profound effect on the physicochemical properties of guest molecules as they are temporarily included within the host cavity.

These properties are:

- solubility enhancement of highly insoluble guests,
- stabilisation of labile guests against the degradative effects of oxidation, visible or UV light and heat,
- control of volatility and sublimation,
- physical isolation of incompatible compounds,
- chromatographic separations,
- taste modification by masking of flavours, unpleasant odours,
- controlled release of drugs and flavours,
- removal of dyes and auxiliaries from dyeing effluents,
- retarding effect in dyeing and finishing,
- protection of dyes from undesired aggregation and adsorption.

2.3. Application Of Cyclodextrins

Complexes can be formed in solutions, in the solid state, as well as when CDs are linked to a solid surface where they can act as permanent or temporary hosts to those small molecules that provide certain desirable attributes such as adsorption of dyestuff molecules, fragrances or antimicrobial agents. This "molecular encapsulation" is already widely utilized in many industrial products, technologies, and analytical methods [7, 18].

Due to the relatively non-polar character of the cavity in comparison to the polar exterior, CD can form inclusion complexes with a wide variety of guest molecules, such as drugs [19, 20, 21], ionic and non-ionic surfactants [23, 24, 25], dyes [26, 27] and polymers [28], etc. The use of CDs has increased annually around 20–30%, of which 80-90% was in food products [29]. In the pharmaceutical industry, CDs and their derivatives have been used either for complexation of drugs or as auxiliary additives such as solubilizers, diluents, or ingredients for improving of drugs physical and chemical properties, or to enhance the bioavailability of poorly soluble moieties [30]. In the chemical industry, CDs and their derivatives are used as catalysts to improve the selectivity of reactions, as well as for the separation and purification of industrial-scale products [31]. In the food, cosmetics, toiletry, and tobacco industries, CDs have been widely used either for stabilization of flavours and fragrances or for the elimination of undesired tastes, microbiological contaminations, and other undesired compounds [7]. For the last 30 years, the use of CDs and their derivatives in the textile domains has captivated a lot of attention. Many of the papers and patents report the use of CDs for finishing and dyeing processes. For instance, they discuss the capture of unpleasant smells due to perspiration, or how to do the controlled release of perfumes, insecticides and antibacterial agents [4, 32-45].

3. CYCLODEXTRINS IN TEXTILES

In the textile field CDs may have many applications such as: they can absorb unpleasant odours; they can complex and release fragrances or “skin-care-active” substances like vitamins, caffeine and menthol as well as bioactive substances such as biocides and insecticides. Further, various textile materials treated with CDs could be used as selective filters for adsorption of small pollutants from waste waters - “preparation of textile nanosponges”.

3.1. Cyclodextrins In Textile Finishing

One of the new concepts for modification of textile substrates is based on the permanent fixation of supramolecular compounds on the material's surface and, thus, imparts new functionality to the fabric. [46]. One of the most promising supramolecular moieties applied to textiles are CDs. Covalent bonding of CDs onto textile fibres was firstly patented in 1980 by Szejtli [47]. He and co-workers reported to bond CD via crosslinking reagent epichlorhydrin onto alkali-swollen cellulose fibres. They found out that CD covalently bonded to cellulose retained the ability to form complexes; when cellulose was treated with a drug it complexed with CD. The complexed drug was upon the contact with the skin released; cellulose textile substrate containing covalently bound β-CD was treated with solution of iodine, potassium iodide and methanol as a solvent to prepare a medical bandage [48].

Szejtli published [4] a very extensive review about CDs in the textile industry. In his review he divided the application of CDs in textile sectors in the following areas: binding of CDs to fibre surfaces, CDs in textile dyeing, in textile finishing, CDs and detergents and miscellaneous applications of CDs in textile industry and textile care. Due to the fact that application of CDs in textile dyeing processes was extensively reported in a chapter of the book edited by Hauser [18], we will emphasis in current publication about the application of CDs in textile finishing.

3.1.1. Binding Of Cyclodextrins To Fibre Surfaces

The attachment of CD molecules on textile substrate provides hosting cavities that can include a large variety of guest molecules for specific functionality. There are two possible approaches to bond CDs onto textile fibres such as chemical bonding of modified CDs on the fibre surfaces or to use bifunctional reagents to link CDs covalently on fibre surfaces.

The most promising approach to bond modified CDs onto textile fibres is the modification of CDs with trichlorotriazines to prepare monochlorotriazinyl-cyclodextrin (CD-MCT) [49, 50]. Analogues to reactive dyes the CD-MCT can be fixed to the fabric by well-known methods and with common equipment. CD-MCT can be applied to fibre surfaces by an exhaustion method or by thermofixation. Moldenhauer with co-workers found out [51] that the fixation was the best when textile substrate was cotton. Mixed fibre materials like cotton/polyurethane or cotton/polyamide can be finished with β-CD-MCT in good yields. Ibrahim et.al [52] reported the improvement of UV protective properties of cotton/wool and viscose/wool blends via incorporating of reactive β-CD-MCT in the easy care finishing formulations, followed by subsequent treatment with copper-acetate or post-dyeing with different classes of dyestuffs (acid, basic, direct and reactive). They found out that post-dyeing of the prefinished textile blends results in a significant increase in the UPF (UV-protection factor) values as a direct consequence of a remarkable reduction in UV radiation transmission through the plain weave fabric. β-CD modified with monochlorotriazine was applied to the cotton fabrics for entrapping of sandalwood oil as an aroma-finishing agent by Sricharussin [53]. The Fourier transform infrared, tensile stress tests and gas chromatography-mass spectroscopy measurements were used to investigate the effects of the treatment. It was found that β-CD-MCT can be fixed to cotton fabrics with the pad-dry-cure method at high temperature. No loss of tensile strength of the treated fabrics was reported. The fragrance disappeared from untreated cotton after 8 days when stored at ambient temperature (30°C) but on other hand, the fragrance was retained in β-CD-MCT-treated cotton fabrics for 21 days in the same conditions. Agrawal et.al compared the efficiency of enzymatic treatments and existing chemical techniques for bonding β-CD and its derivatives to cotton surface. Novel chemical based crosslinking with homo-bi-functional reactive dye (C.I. reactive black 5) and grafting with reactive β-CD-MCT show maximum attachment to cotton surface. Innovative, enzymatic coupling of especially synthesized 6-monodeoxy-6-mono(N-tyrosinyl)-β-cyclodextrin was performed on cotton textile surface at low temperature. Alteration in surface topography has been observed for all β-CD treated samples [54]. Martel with co-workers coupled β-CD-MCT to chitosan, to obtain a chitosan derivative bearing cyclodextrin. Because the average degree of substitution of the CD derivative was 2.8, the reaction yielded crosslinked insoluble products. The structure of these materials has been investigated by high-resolution magic-angle spinning (HRMAS) with gradients. For the first time, HRMAS spectra of chitosan polymers containing β-CD were obtained. This NMR technique produced one- and two-dimensional well-resolved solid-state spectra. Decontamination of waters containing textile dyes were carried out with the crosslinked derivatives. Report by Martel showed that the new chitosan derivatives are characterized by a rate of sorption and a global efficiency superior to that of the parent chitosan polymer and of the well-known cyclodextrin-epichlorohydrin gels [55]. El-Tahlawy with co-workers carried out a novel technique for preparation of cyclodextrin-grafted chitosan. β-CD citrate was synthetized by esterifying of β-CD with citric acid (CA) in presence or absence of sodium hypophosphite as a catalyst in a semidry process. β-CD/grafted chitosan was prepared by coupling β-CD citrate with chitosan dissolved in different formic acid solutions having different concentrations. The reacting ingredients were subjected to various reaction conditions to attain the optimum condition. β-CD/grafted chitosan were evaluated by measuring the nitrogen content of both chitosan and grafted chitosan. Chitosan and β-CD/grafted chitosan, having different molecular weights, were evaluated as antimicrobial agents for different microorganisms [56].

Very effective bonding of CDs on cellulose fibres can be achieved by a high-performance resin finish [57] or with non-formaldehyde reagents such as polycarboxylic acids [58, 59] which can covalently esterify hydroxyl groups of cellulose and CDs and therefore the grafting of CD on cellulose can occur. The same linking/crosslinking reagents can be used in the treatment of different synthetic fibres. Polyester fibres were modified by β-CD using citric acid [7, 59] in research work of our group [60], 1,2,3,4-butane tetracarboxylic acid was used as a linker. Odour control is a very important topic in the apparel and underwear items. Odour can be controlled by applying an antimicrobial finish, removing the odour molecules as they are formed or covering up the odour with a fragrance. The odour molecules being hydrophobic become trapped in the cavities of the CDs and are removed during laundering.

Within our research work PET fibres were treated in aqueous solution of different concentrations of β–CD and BTCA. We reported that BTCA molecules react via anhydride formation with hydroxyl groups of β-CD and form nano-assembly which can be physically attached to the PET fibres surface at the elevated temperature. Such assembly could be schematically presented as shown in Figure 3.

For reducing the termofixation temperature the catalyst cyanamide was used. We concluded that the treatment of PET with β-CD/BTCA was very successful even at temperature as low as 115°C when CA as a catalyst has been used. After 10 washings the gain on mass remained as high as 7.8% (Figure 4).

We were able to reduce the curing temperature to 115ºC and prepared nanoencapsulated textile materials with increased adsorption capacity (the adsorption of ammonia gas onto treated and untreated PET textile materials was measured using Japan standard test method - JIS K0804) and with postponed release of volatile compounds. From the ammonia gas adsorption measurements (Table 1), it is possible to conclude that the adsorption of ammonium gas increased when PET fabric was treated with β-CD/BTCA/CA at 115ºC/10min - after one hour of exposure to ammonia gas the concentration of gas in the chamber was zero, compare to the concentration when untreated PET fabric was exposed to the ammonium gas, where the concentration in the chamber was changed from the initial value of 125ppm to 77ppm.

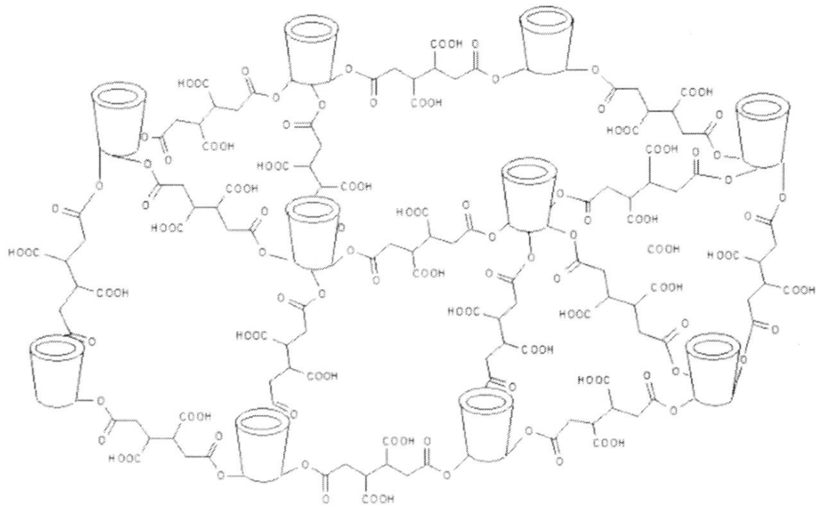

Figure 3. Nano-assembly of β-CD crosslinked with BTCA on textile surface.

In order to quantify the odour-releasing behaviour of β-CD treated PET fabrics, we organized a sensory panel of nine people to whom the odour was presented under controlled conditions. In order to study the postponed release of the volatile compounds from the β-CD treated textile substrate the following was performed: β-CD/BTCA/CA treated PET textile substrate was sprayed with perfume and dried; the intensity of the perfume from the untreated PET fabric spayed with perfume was also monitored for comparison purposes. Both treatments were performed in triplets. The size of the clothes was 10 by 10 cm. All with perfume treated textile samples were stored separately in dark places. Samples were stored in open conditions so that the perfume was able to evaporate constantly. The odour release was measured once per week. The smell intensity was evaluated from 0 to 4, where 0 means no smell and 4 means very intensive smell. From Figure 5 it is possible to see that the odour release intensity of untreated PET fabrics sprayed with perfume (BLIND, spray) starts to decrease after 6 weeks; the odour intensity of perfume sprayed on β-CD treated fabrics remains constant, but there is slight indication that the intensity of the perfume starts to increase after 6 weeks. We can conclude that, in the case of β-CD treated PET fabrics, some postponed release of the fragrance occurs.

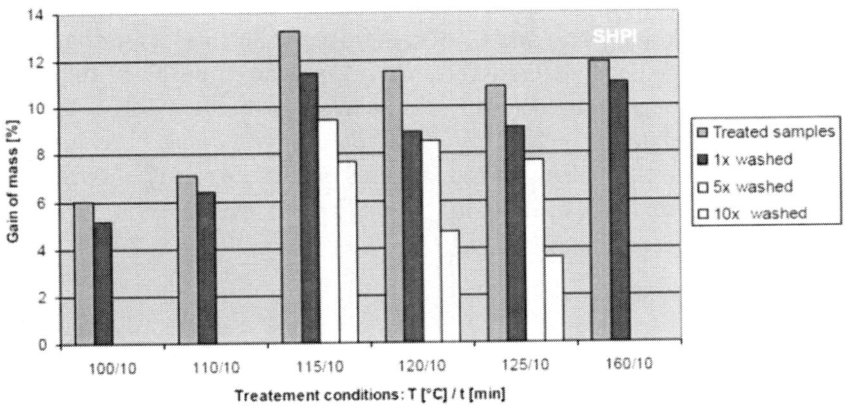

Figure 4. Samples 100/10, 110/10, 115/10, 120/10 and 125/10 are PET samples treated with β-CD, BTCA, CA at 100, 110, 115, 120 and 125°C, respectively; sample 160/10 was treated with β-CD, BTCA and SHPI at 160°C; all samples were termofixed for 10 minutes.

Table 1. Decrease of ammonium gas concentration due to the adsorption.

	PET treated with β-CD/BTCA/ CA at 115°C/10 min	Untreated PET
Initial concentration (ammonia)	125ppm	125ppm
One hour concentration (ammonia)	0ppm	77ppm

Glycidyl methacrylate is widely used in the production of polymer coatings and finishes, adhesives, plastics and elastomers, it can be grafted to various textiles substrates as well. Desmet and co-workers functionalized cotton-cellulose by gamma-irradiation-induced grafting of glycidyl methacrylate (GMA) to obtain a hydrophobic cellulose derivative with epoxy groups suitable for further chemical modification. Two grafting techniques were applied. In pre-irradiation grafting (PIG) cellulose was irradiated in air and then immersed in a GMA monomer solution, whereas in simultaneous grafting (SG) cellulose was irradiated in an inert atmosphere in the presence of the monomer. In the paper authors claimed that the PIG led to a more homogeneous fibre surface, while SG resulted in higher grafting yield but showed clear indications of some GMA-homopolymerization. Effects of the reaction parameters (grafting method, absorbed dose, monomer concentration, solvent composition) were evaluated by SEM, gravimetry (grafting yield) and FT-IR spectroscopy. It is reported that water uptake of the cellulose decreased while adsorption of pesticide molecules increased upon grafting. The adsorption was further enhanced by β-CD immobilization during SG. This method can be applied to produce adsorbents from cellulose based agricultural wastes [61]. CDs can be incorporated into fibres during the spinning processes [62, 63].

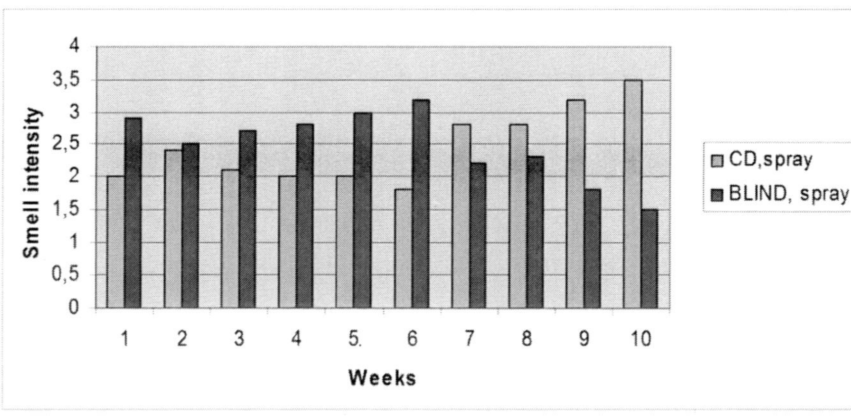

Figure 5. Odour intensity of PET fabrics pre-treated with β-CD (blue) and untreated PET fabrics sprayed with the perfume.

Electrospinning is proven to be an effective method for producing non-woven mats of fibres with high aspect ratios. Manasco and co-workers reported the preparation of submicron hydroxypropyl-beta-cyclodextrin (HP-β-CD) fibres by electrospinning without the addition of a carrier polymer. They focused on exploring solution properties that make fibre formation possible contrary to the widely accepted premise that molecular entanglement of macromolecules is required for electrospinning. The ability to electrospin from these solutions was attributed to hydrogen-bonded aggregation between HP-β-CD molecules at high concentrations [64]. Further it is reported by Uyar and co-workers that poly(methyl methacrylate) (PMMA) nanofibres containing the inclusion complex forming β-CD were successfully produced by means of electrospinning in order to develop functional nanofibrous webs. Electrospinning of uniform PMMA nanofibres containing different loadings of β-CD (10%, 25% and 50% (w/w)) was achieved. The surface sensitive spectroscopic techniques; X-ray photoelectron spectroscopy (XPS) and time-of-flight secondary ion mass spectrometry (ToF-SIMS) showed that some of the β-CD molecules are present on the surface of the PMMA nanofibres, which is essential for the trapping of organic vapours by inclusion complexation. Direct pyrolysis mass spectrometry (DP-MS) studies showed that PMMA nanowebs containing β-CD can entrap organic vapours such as aniline, styrene and toluene from the surroundings due to inclusion complexation with beta-CD that is present on the fibre surface. The study showed that electrospun nanowebs functionalized with CDs may have the potential to be used as molecular filters and/or nanofilters for the treatment of organic vapour waste and air filtration purposes [65]. Polyvinyl alcohol (PVA) nanowebs incorporating vanillin/cyclodextrin inclusion complex (vanillin/CD-IC) were produced via electrospinning technique by Kayaci and Uyar [66]. The vanillin/CD-IC was prepared with three types of CDs; α-CD, β-CD and γ-CD to find out the most favourable CD type for the stabilization of vanillin. PVA/vanillin/CD-IC nanofibres, having fibre diameters similar to 200 nm, were electrospun from aqueous mixture of PVA and vanillin/CD-IC. The results indicated that vanillin with enhanced durability and high temperature stability was achieved for PVA/vanillin/CD-IC nanowebs due to complexation of vanillin with CD. Additionally, they reported that PVA/vanillin/γ-CD-IC nanoweb was more effective for the stabilization and slow release of vanillin suggesting that the strength of interaction between vanillin and the γ-CD cavity is stronger when compared to α-CD and β-CD.

3.1.2. Cosmetotextiles

Although still in its infancy, the market for cosmetotextiles - often referred to as "wearable skincare" - is set to grow rapidly, and the textile industry is optimistic that further technical developments will open up new markets and create growing business opportunities [67]. Cosmetotextile became a fast growing new branch of specialized micro- or nano-encapsulation textile products, with most patents issued after the 1990 [68]. The physical and chemical properties of the guest molecules encapsulated in CDs can change due to complex formation. Thus, for example, the stability of the complexed molecule against light and oxygen increases and the vapour pressure is reduced. The solubility of slightly soluble molecules increases in a CD complex. All these and further advantages of CDs and their complexes can be used for the formulation of cosmetic products [69]. Cosmetotextile allows the administration of active molecules simply and controllably. It can also be used to change the surface properties of a fabric in order to make it self-cleaning, hydrophobic or lipophobic. The article prepared by Ripoll and co-workers reviews the current state of the art concerning functionalization techniques and the methods used to characterize various functionalized fabric. This review also reveals the surprising lack of publications on the functionalization of textile supports [70].

Moist oils (essential oils, herbal oils, oils from flower seeds) have skin care benefits in that they provide an occlusive layer that lubricates the epidermis, together with a moisturizing effect that helps to prevent excess water loss. Essential oils attributed with a range of properties that help to achieve physical and emotional balance. Besides, one additional advantage of molecular encapsulation is the possibility to reload them 4, 58, 71. Cosmetotextile applications can be used in the treatment of chronic venous insufficiency in legs by means of elastic bandages loaded with natural products which possess flebotonic properties. Cravotto and co-workers 72 have developed an efficient synthetic procedure for the preparation of β-CD-grafted viscose by means of the 2-step ultrasound-assisted reaction. The highly grafted fabric bearing bis-urethane bridged β-CD has been characterized by ATR FT-IR and CP-MAS spectra and by an empiric colorimetric method which used phenolphthalein as the CD guest. They have also developed a suitable cosmetic preparation containing natural substances and extracts (aescin, menthol, Centella asiatica and Ginkgo biloba) to recharge the CD-grafted textile. The efficacy of the new cosmetotextile has been corroborated by in vitro studies of diffusion through membranes, cutaneous permeation and accumulation in porcine skin. Aescin was taken as a reference compound and its concentration in the different compartments was monitored by HPLC analysis. They reported that this cost effective cosmetotextile showed excellent application compliance and was easily recharged.

Electrospun functional polystyrene (PS) fibres containing CD-menthol inclusion complexes are new and advance class of cosmetotextiles. Uyar and co-workers developed functional electrospun fibres containing fragrances/flavours with enhanced durability and stability assisted by CD inclusion complexation. As a model fragrance/flavour molecule they used menthol. CD-menthol inclusion complexes were incorporated in electrospun PS fibres by using three types of CDs: α-CD, β-CD and γ-CD. It is reported that due to complexation of menthol with CDs, the stabilization of menthol was achieved up to 350°C whereas the PS fibres without the CD complex could not preserve volatile menthol molecules. This study suggested that the electrospun fibres functionalized with CD are very effective for enhancing the temperature stability of volatile fragrances/flavours and therefore show potentials for the development of functional fibrous materials [73].

Today the subject of well-being is an area which is receiving much interest, with scent being one of the most important aspects of personal care. The definition of the word aromatherapy is the following: therapeutic uses of fragrances which at least mere volatilize to cure and to mitigate or cure diseases, infection and indisposition by means of inhalation alone [74]. The term aromachology was coined in 1982 to denote the science that is dedicated to the study of the relationship between psychology and fragrance technology to elicit a variety of specific feelings and emotions – such as relaxation, exhilaration, sensuality, happiness and well-being-through odours.

CDs linked on cellulose do not affect the cellulose's properties, and CDs keep their ability to form inclusion complexes with other suitable molecules thus, CDs are the first choice in preparing aromatherapy textiles. Lavender is the most used and most versatile of all the essential oils. It is very useful oil, especially when symptoms are due to a nervous problem. The effects of lemon, camomile, rose, cardamom, clove, and jasmine fragrance oils on human have been confirmed by many research works. The sedative effects for the pharmaceutical and emotional effects of essential oils are listed inTables 2 and 3 respectively [75, 76].

3.1.3. Miscellaneous Applications Of Cyclodextrins In Textile Industry

An insect repellent is a substance applied to skin, clothing, or other surfaces which discourages insects from landing or climbing on that surface. Synthetic repellents (such as paradichlorobenzene) tend to be more effective than 'natural' repellents, but on the other hand they are usually toxic. Cedar oil is often used as a natural insect repellent or it is used for its aromatic properties, especially in aromatherapy. Essential oil repellents tend to be short-lived in their effectiveness due to their volatile nature. To prevent the essential oils from evaporating form the textile materials we can encapsulate them in β-CD. In our research group [80, 81] with β-CD, nanoencapsulated wool and PET/wool blend fibres were further treated with cedar oil, which is known for being a natural insect repellent. The complex formation of cedar oil with β-CD was determined by ATR FT-IR spectroscopy. Textile material containing β-CD showed, after being treated with cedar oil, a prolonged moths oppression compared to those textile materials treated only with cedar oil. Table 4 presents the damages to wool and larval conditions according to the time of exposure. No visible damage was observed to the naked eye when β-CD/cedar oil treated wool was exposed to a moth's colony for 2 months. In the control (wool samples treated only with cedar oil) no damage was observed for the first few days, but when the cedar oil had evaporated, the wool cloth was not protected anymore. In contrast, when the cedar oil was encapsulated in the β-CD cavity, evaporation was hindered and resistance to insect pests' activities regarding cedar oil remained.

To maintain antimicrobial activity, frequent administration of conventional formulations of many antibiotics with short half-life is necessary. To enhance release properties, many materials have been introduced into the matrix and coating extended-release system in the past few years. The review by Gao [82] highlights the development of materials used in extended-release formulation and nanoparticles for antibiotic delivery. CDs are mentioned as nanoparticles/nanocarriers which allow the antibiotic extended-release.

A textile polyester vascular graft can be modified by methyl-β-CD to obtain a new implant capable of releasing antibiotics directly in situ at the site of operation over a prolonged period and thereby preventing post-operative infections [83]. Wang reported the inclusion complex of miconazole nitrate with β-CD formation by the co-precipitation method. The DSC curve and X-ray diffraction verified the inclusion complex formation between β-CD and miconazole nitrate. The skin-care textiles can be obtained by treating fabrics with inclusion complexes using the sol-gel method [84].

Table 2. Pharmaceutical effect of essential oils [**75, 77, 78**].

Effects	Essential Oil
Sedation	Mint, Onion, Lemon, Metasequoia
Coalescence	Pine, Clove, Lavender, Onion, Thyme
Diuresis	Pine, Lavender Onion. Thyme, Fennel, Lemon, Metasequoia
Facilitating Menses	Pine, Lavender. Mint, Rosemary, Thyme, Basil, Chamomile, Cinnamon, Lemon
Dismissing sputum	Onion, Citrus, Thyme, Chamomile
Allaying a fever	Ginger, Fennel, Chamomile, Lemon
Hypnogenesis	Lavender, Oregano, Basil, Chamomile
Curing Hypertension	Lavender, Fennel, Lemon, Ylangylang
Be good for stomach	Pine, Ginger, Clove, Mint, Onion, Citrus, Rosemary, Thyme, Fennel, Basil, Cinnamon
Diaphoresis	Pine, Lavender, Rosemary, Thyme, Chamomile, Metasequoia
Expelling wind	Ginger, Clove, Onion, Citrus, Rosemary, Fennel, Lemon
Losing weigh	Onion, Cinnamon, Lemon
Relieving pain	Vanilla, Lavender. Mint, Onion, Citrus, Rosemary, Chamomile, Cinnamon, Lemon
Detoxification	Lavender
Curing diabetes	Vanilla, Onion, Chamomile, Lemon
Stopping diarrhea	Vanilla, Ginger, Clove, Lavender, Mint, Onion, Oregano, Rosemary, Thyme, Chamomile, Cinnamon, Lemon
Curing flu	Pine, Lavender, Mint, Onion, Citrus, Rosemary, Thyme, Chamomile, Cinnamon, Metasequoia
Curing rheumatism	Lavender, Onion, Citrus, Rosemary, Thyme, Metasequoia
Urging sexual passion	Pine, Ginger, Clove, Mint, Onion, Rosemary, Thyme, Fennel
Relieving spasm	Cinnamon
Promoting appetite	Clove, Lavender, Mint, Onion, Citrus, Rosemary, Fennel, Basil, Chamomile, Cinnamon, Lemon, Metasequoia
Relieving cough	Rosemary

Novel nano-porous polymers or nanosponges can be prepared for removal of organic pollutants from waste water. The polymeric «nanosponge» materials are not durable (usually they are in gel form), they do not have high mechanical strength, so they must be impregnated onto the pore structure of a ceramic or some other porous surfaces [85, 86]. This technology is very specific for the target pollutant, it is very expensive and the removal of the adsorbed pollutant from the nanosponge is not possible. Textile materials are very important as filter materials. The cost of textile materials is acceptable (polyester, viscose), they have sufficient mechanical strength; the pore size, especially the macro-pore size can vary and it depends on the type of textile (the density of non-woven material) and on the diameter of the fibres. Textile materials can be further modified to prepare filtration materials with additional adsorption.

Table 3. The sedative or emotion effects of essential oils [**75**, **79**].

Anxiety	Benzoin, Lemon, Chamomile, Rose, Cardamom, Clove, Jasmine
Lament	Rose
Stimulation	Camphor, Balm oil
Anger	Chamomile, Balm oil, Rose, Ylangylang
Wretchedness	Basil, Cypress, Mint, Patchouli
Allergy	Chamomile, Jasmine, Balm oil
Distrustfulness	Lavender
Tension	Camphor, Cypress, Vanilla, Jasmine, Balm oil, Lavender, Sandalwood
Melancholy	Basil, Lemon, Chamomile, Vanilla, Jasmine, Lavender, Mint, Rose
Hysteria	Chamomile, Balm oil, Lavender, Jasmine
Mania	Basil, Jasmine, Pine
Irritability	Chamomile, Camphor, Cypress, Lavender
Desolation	Jasmine, Pine, Patchouli, Rosemary

Table 4. The estimation of wool damage and the condition of larval colony in accordance to elapsed time.

Time	β-CD/cedar oil treated wool	Cedar oil treated wool	Untreated wool
48h	No detectable damage	No detectable damage	Very slight visible damage
	Larval conditions: live	Larval conditions: live	Larval conditions: live
72h	No detectable damage	No detectable damage	Moderate visible damage
	Larval conditions: dead	Larval conditions: dead	Larval conditions: live
7 days	Addition of new larvae	Addition of new larvae	/
14 days	No detectable damage	Very slight visible damage	Very heavy damages
	Larval conditions: dead	Larval conditions: live	Larval conditions: live
56 days	No detectable damage	Moderate visible damages	Very heavy damages
	Larval conditions: dead	Larval conditions: live	Larval conditions: live, pupating

The amount of aromatic organic pollutants (phenols, aniline, formaldehyde and others) can be reduced from dyeing wastewater by using CDs which can be immobilized on a water insoluble organic support. The new concept for modification of textile substrates based on permanent fixation of supramolecular compounds - CDs on the material surface thus imparts new functionality to the fabric [87]. The guest molecules could be various organic molecules and some metal ions as well. The assembly of nanocapsules on textile materials acts as selective filtration/adsorption media for various pollutants. Prabaharan and Mano further reported that CDs have recently been recognized as useful adsorbent matrices. Due to its hydrophobic cavity, CDs can interact with appropriately sized molecules to result in the formation of inclusion complexes. These complexes are of interest for scientific research as they exist in aqueous solution and can be used to study the hydrophobic interactions which are important in the biomedical and environmental fields. The grafting of CD onto chitosan can result in the formation of a molecular carrier that possess the cumulative effects of inclusion, size specificity and transport properties of CDs as well as the controlled release ability of the polymeric matrix. In this review, different methods of CD grafting onto chitosan are discussed [88]. Electrospinning has been used to create polystyrene (PS) nanofibres containing any of the three different types of cyclodextrin (CD); α-CD, β-CD, and γ-CD [89]. These three CDs are chosen because they have different sized cavities that potentially allow for selective inclusion complex (IC) formation with molecules of different sizes or differences in affinity of IC formation with one type of molecule. The comparative efficiency of the PS/CD nanofibres/nanoweb for removing phenolphthalein, a model organic compound, from solution was determined by UV-Vis spectrometry, and the kinetics of phenolphthalein capture was shown to follow the trend PS/α-CD > PS/β-CD > PS/γ-CD. Direct pyrolysis mass spectrometry (DP-MS) was also performed to ascertain the relative binding strengths of the phenolphthalein for the CD cavities, and the results showed the trend in the interaction strength was β-CD > γ-CD > α-CD. Results of their research demonstrated that nanofibres produced by electrospinning that incorporate CDs with different sized cavities can indeed filter organic molecules and can potentially be used for filtration, purification, and/or separation processes.

4. CONCLUSION

Since 1980, when Szejtli first patented the bonding of CDs onto textile fibres, a lot of research has gone into the application of CDs on to textile substrates. But there is still a gap between original high level basic science and commercial applications of CDs in all industrial sectors.

Nevertheless the use of CDs in the textile industry has increased in the last years. Grafted CDs on textile substrates or spun fibres which contain CDs can be used to obtain special functionality of textiles such as absorption; they can complex and release fragrances or "skin-care-active" substances like vitamins, caffeine and menthol as well as bioactive substances such as biocides and insecticides and drugs. Furthermore, various textile materials treated with CDs could be used for adsorption of small pollutants from waste waters - for filtration, purification, and/or separation treatments of waste waters.

REFERENCES

1. Vögtle F. Supramolecular Chemistry, an introduction. New York: John Wiley & Sons; 1991.
2. Astray G, Gonzalez-Barreiro C, Mejuto JC, Rial-Otero R, Simal- Gándara J. A review on the use of cyclodextrins in foods. Food Hydrocolloids 2009;23(7) 1631–1640.
3. Jeang CL, Lin DG, Hsieh SH. Characterization of cyclodextrin glycosyltransferase of the same gene expressed from Bacillus macerans, Bacillus subtilis, and Escherichia coli. Journal of Agricultural and Food Chemistry 2005;53(16) 6301-6304.
4. Szejtli J. Cyclodextrins in the Textile Industry. Starch/Stärke 2003;55(5) 191-196.
5. Cravotto G, Binello A, Baranelli E, Carraro P, Trotta F. Cyclodextrins as food additives and in food processing. Current Nutrition & Food Science 2006;2(4) 343–350.
6. Dastjerdi R, Montazer M. A review on the application of inorganic nano-structured materials in the modification of textiles: Focus on anti-

microbial properties. Colloids and Surfaces B: Biointerfaces 2010;79(1) 5-18.

7. Martin del Valle EM. Cyclodextrins and their uses: a review. Process Biochemistry 2004;39(9) 1033–1046.

8. Li S, Purdy WC. Cyclodextrins and their Applications in Analytical-Chemistry. Chemical Reviews 1992;92(6) 1457-1470.

9. Szejtli J. Introduction and General Overview of Cyclodextrin Chemistry. Chemical Reviews 1998;98(5) 1743–1753.

10. Uekama K, Hirayama F, Irie T. Cyclodextrin Drug Carrier Systems. Chemical Reviews 1998;98(5) 2045-2076.

11. Ribeiro ACF, Valente AJM, Lobo VMM. Transport Properties of Cyclodextrins: Intermolecular Diffusion Coefficients. Journal of the Balkan Tribological Association 2008;14(3) 396-404.

12. Harada Laboratory. Department of Macromolecular Science. Osaka University. http://www.chem.sci.osaka-u.ac.jp/lab/harada/eng/eng/research/01.html (accessed 2 July 2012)

13. Saenger W. Sterochemistry of circularly closed oligosaccharides: cyclodextrins structure and function. Biochemical Society Transactions 1983;11(2) 136-139 Connors, K. A. (1997). The Stability of Cyclodextrin Complexes in Solution. Chemical Reviews, Vol.97, No.5, 325-1357, ISSN: 0009-2665

14. Connors KA. The Stability of Cyclodextrin Complexes in Solution. Chemical Reviews 1997;97(5) 325-1357.

15. Weber E. Molecular Inclusion and Molecular Recognition - Clathrates I. Topics in Current Chemistry 1987;140 1-20.

16. Montazer M, Jolaei MM. β-Cyclodextrin stabilized on three-dimensional polyester fabric with different crosslinking agents. Journal of Applied Polymer Science 2010;116(1) 210–217.

17. Lo Meo P, D'Anna F, Riela S, Gruttadauria M, Noto R. Spectrophotometric study on the thermodynamics of binding of α- and β-cyclodextrin towards some p-nitrobenzene derivatives. Organic & Biomolecular Chemistry 2003;1(9) 1584-1590.

18. Voncina B. Application of cyclodextrins in textile dyeing. In: Hauser PJ (ed.) Textile dyeing. Rijeka: InTech; 2011. p373-392.

19. Fernandes CM, Carvalho RA, da Costa SP, Veiga FJB. Multimodal molecular encapsulation of nicardipine hydrochloride by β-cyclodextrin, hydroxypropyl-β-cyclodextrin and triacetyl-β-cyclodextrin in solution. Structural studies by 1H NMR and ROESY experiments. European Journal of Pharmaceutical Sciences 2003;18(5) 285-296.

20. Figueiras A, Sarraguca JMG, Carvalho RA, Pais AACC, Veiga FJB. Interaction of Omeprazole with a Methylated Derivative of β-Cyclodextrin: Phase Solubility, NMR Spectroscopy and Molecular Simulation. Pharmaceutical Research 2007;24(2) 377-389.

21. Monti S, Sortino S. Photoprocesses of photosensitizing drugs within cyclodextrin cavities. Chemical Society Reviews 2002;31(5) 287-300.

22. Nilsson M, Cabaleiro-Lago C, Valente AJM, Soderman O. Interactions between gemini surfactants, 12-s-12, and beta-cyclodextrin as investigated by NMR diffusometry and electric conductometry. Langmuir 2006;22(21) 8663-8669.

23. Cabaleiro-Lago C, Nilsson M, Soderman O. Self-diffusion NMR studies of the host-guest interaction between beta-cyclodextrin and alkyltrimethylammonium bromide surfactants. Langmuir 2005;21(25) 11637-11644.

24. Valente AJM, Nilsson M, Soderman O. Interactions between n-octyl and n-nonyl β-d-glucosides and α- and β-cyclodextrins as seen by self-diffusion NMR. Journal of Colloid and Interface Science 2005;281(1) 218-224.

25. Garcia-Rio L, Godoy A. Use of spectra resolution methodology to investigate surfactant/β-cyclodextrin mixed systems. Journal of Physical Chemistry B 2007;111(23) 6400-6409.

26. Voncina B, Vivod V, Jausovec D. [Beta]-cyclodextrin as a retarding reagent in polyacrylonitrile dyeing. Dyes and pigments 2007;74(3) 642-646.

27. Costa T, Seixas de Melo JS. The effect of γ-cyclodextrin addition in the self-assembly behavior of pyrene labeled poly(acrylic) acid with

different chain sizes. Journal of Polymer Science Part a-Polymer Chemistry 2008;46(4) 1402-1415.

28. Denter U, Schollmeyer E. Surface modification of synthetic and natural fibres by fixation of cyclodextrin derivatives. Journal of Inclusion Phenomena and Molecular Recognition in Chemistry 1996;25(1-3) 197-202.

29. Szejtli J. Utilization of cyclodextrins in industrial products and processes. Journal of Materials Chemistry 1997;7(4) 575–587.

30. Frömming KH, Szejtli, J. Cyclodextrins in Pharmacy. Dordrecht: Kluwer Academic Publishers; 1994.

31. Hedges AR. Industrial applications of cyclodextrins. Chemical Reviews 1998;98(5) 2035-2044.

32. Buschmann HJ, Knittel D, Schollmeyer E. Textile Materialien Mit Cyclodextrinen, Textile Materials with Cyclodextrins. Melliand Textilberichte 2001;82(5) 368-370.

33. Buschmann HJ. Cosmetic textiles: Clothes with a skin care action. Cossma 2001;2(7) 38-39.

34. Buschmann HJ, Knittel D, Beermann K, Schollmeyer E. Cyclodextrins and textiles. Nachr. Chem. 2001;49(5) 620-620.

35. Murthy CN, Shown I. Grafting of Cotton Fiber by Water-Soluble Cyclodextrin-Based Polymer. Journal of Applied Polymer Science 2009;111(4) 2056-2061.

36. Savarino P, Viscardi G, Quagliotto P, Montoneri E, Barni E. Reactivity and effects of cyclodextrins in textile dyeing. Dyes and Pigments, 1999;42(2) 143-147,.

37. Poulakis, K, Buschmann HJ, Schollmeyer E. Ger. Offen. 4035378. 1992; WO 2002046520. 2002.

38. Buschmann HJ, Knittel D, Schollmeyer E. Resin finishing of cotton in the presence of cyclodextrins for depositing fragrances. Melliand Textilberichte 1991;72(3) 198-199.

39. Fujimura T. Jpn. Kokai Tokkyo Koho 60259648. 1985.

40. Ritter W, Delney J, Volz W, Kerr I. DE 10101294. 2002.

41. Akasaka M, Shibata T, Ochia H. Jpn. Kokai Tokkyo Koho 03059178. 1991.

42. Akasaka, M, Sawai Y, Iwase K, Moriishi H. EP 488294. 1992.

43. Yamamoto K, Saeki T. Jpn. Kokai Tokkyo Koho 09228144. 1997.

44. Yamamoto K, Saeki T. Jpn. Kokai Tokkyo Koho 10007591. 1998.

45. Voncina B, Majcen N. Application of cyclodextrin for medical and hygienic textiles. Tekstil 2004;53(1) 1-9.

46. Knittel D, Schollmeyer E. Technologies for a new century. Surface modification of fibres. Journal of the Textile Institute 2000;91(3) 151-165.

47. Szejtli J, Zsadon B, Fenyvesi E, Otta K, Tudos F. Hungarian Patent: 181733. 1980. US Patent 4,357,468 (1982).

48. Szejtli J, Zsadon B, Horvath OK, Ujhazy A, Fenyvesi E. Hungarian Patent: 54506. 1991.

49. Reuscher H, Hirsenkorn R, Haas W. German Patent: DE 19520967. 1996.

50. Grechin AG, Buschmann HJ, Schollmeyer E. Quantification of Cyclodextrins fixed onto Cellulose Fibres. Textile Research Journal 2007;77(3) 161-164.

51. Moldenhauer JP, Reuscher H. Textile finishing with MCT-beta-cyclodextrin. In: Torres Labandeira JJ, Vila-Jato JL. (eds.) Proceedings of the 9th International Symposium on Cyclodextrins, 31 May – 3 June 1998, Santiago de Compostela, Spain. Dordrecht: Kluwer Academic Publishers; 1999.

52. Ibrahim NA, Allam EA, El-Hossamy MB, El-Zairy WM. UV-Protective Finishing of Cellulose/Wool Blended Fabrics. Polymer-Plastics Technology and Engineering 2007;46(9), 905-911.

53. Sricharussin W, Sopajaree C, Maneerung T, Sangsuriya N. Modification of cotton fabrics with beta-cyclodextrin derivative for aroma finishing. Journal of the Textile Institute 2009;100(8) 682-687.

54. Agrawal PB, Warmoeskerken MMCG. Permanent fixation of β-cyclodextrin on cotton surface: An assessment between innovative and

established approaches. Journal of Applied Polymer Science 2012;124(5) 4090-4097.

55. Martel B, Devassine M, Crini G, Weltrowski M, Bourdonneau M, Morcellet M. Preparation and sorption properties of a beta-cyclodextrin-linked chitosan derivative. Journal of Polymer Science Part A-Polymer Chemistry 2001;39(1) 169-176.

56. El-Tahlawy K, Gaffar MA, . Novel method for preparation of beta-El-Rafie Scyclodextrin/grafted chitosan and it's application. Carbohydrate Polymers 2006;63(3) 385-392.

57. Ostertag H. Anwendung von β-Cyclodextrinen in der CO-Gewebeveredlung. Melliand Textilberichte 2002;83(11-12) 872-878.

58. Voncina B, Le Marechal AM. Grafting of cotton with beta-cyclodextrin via poly(carboxylic acid). Journal Of Applied Polymer Science 2005; 96(4) 1323-1328.

59. Martel B, Weltrowski M, Ruffin D, Morcellet M. Polycarboxylic acids as crosslinking agents for grafting cyclodextrins onto cotton or wool fabrics: Study of the process parameters. Journal of Applied Polymer Science 2002;83(7) 1449-1456.

60. Vončina B, Le Marechal AM. [Beta]-cyclodextrin in medical and hygienic textiles. In: Chen X, Ge Y, Yan X. (eds.). 83rd TIWC. Quality textiles for quality life: proceedings of the Textile Institute 83rd World Conference (83rd TIWC), 23-27 May 2004, Shanghai, China. Manchester: Shanghai: Textile Institute, Donghua University; 2004.

61. Desmet G, Takacs E, Wojnarovits L, Borsa J. Cellulose functionalization via high-energy irradiation-initiated grafting of glycidyl methacrylate and cyclodextrin immobilization. Radiation Physics and Chemistry 2011;80(12) 1358-1362.

62. He Y, Inoue Y. Effect of alpha-cyclodextrin on the crystallization of poly(3-hydroxybutyrate). Journal of Polymer Science Part B-Polymer Physics 2004;42(18) 3461-3469.

63. Vogel R, Tandler B, Haussler L, Jehnichen D, Brunig H. Melt spinning of poly(3-hydroxybutyrate) fibers for tissue engineering using alpha-cyclodextrin/polymer inclusion complexes as the nucleation agent. Macromolecular Bioscience 2006;6(9) 730-736.

64. Manasco JL, Saquing CD,Tang C,Khan SACyclodextrin fibers via polymer-free electrospinning. RSC Advances 2012;2(9) 3778-3784

65. Uyar T, Havelund R,Nur Y,Balan A , Hacaloglu JToppare L, Besenbacher F,Kingshott P. functionalized poly(methyl methacrylate) (PMMA) electrospun nanofibers for organic vapors waste treatment. Journal of Membrane Science 2010;365(1-2) 409-417.

66. Kayaci F, Uyar T. Encapsulation of vanillin/cyclodextrin inclusion complex in electrospun polyvinyl alcohol (PVA) nanowebs: Prolonged shelf-life and high temperature stability of vanillin Food Chemistry 2012;133(3) 641-649

67. Hague L. Cosmetotextiles market has exciting potential. Aroq Ltd http://www.just-style.com/analysis/cosmetotextiles-market-has-exciting-potential_id111926.aspx (accessed 3 July 2012).

68. CEN standard CEN/TC 248 Textiles and textile products; WG 25 Cosmeto-textiles. CEN - ISO standardization committees on textiles Cheng S.Y. "Development of Cosmetic Texiles Using Microencapsulation Technology", RJTA, vol.12, no.4, 2008

69. Buschmann HJ, Schollmeyer E. Applications of cyclodextrins in cosmetic products. Journal of Cosmetic Science 2002;53(3) 185-191.

70. Ripoll L, Bordes C, Etheve S, Elaissari A, Fessi H. Cosmeto-textile from formulation to characterization: an overview. e-Polymers 2010;040 http://www.e-polymers.org/journal/papers/hfessi_010410.pdf (accessed 5 July 2012).

71. Buschmann HJ, Knittel D, Schollmeyer E. New Textile Applications of Cyclodextrins. Journal of Inclusion Phenomena and Macrocyclic Chemistry 2001;40(3) 169-172.

72. Cravotto G, Beltramo L, Sapino S, Binello A, Carlotti ME. A new cyclodextrin-grafted viscose loaded with aescin formulations for a cosmeto-textile approach to chronic venous insufficiency. Journal of Materials Science-Materials in Medicine 2011;22(10) 2387-2395

73. Uyar T, ,Hacaloglu J .Besenbacher F Electrospun polystyrene fibers containing high temperature stable volatile fragrance/flavor facilitated by cyclodextrin inclusion complexes. Reactive & Functional Polymers 2009;69(3) 145-150.

74. Buchbauer G. Aromatherapy: Use of fragrance and essential oils as medicaments. Flavour and Fragrance Journal 1994;9 217-222.

75. Wang CX, Chen SH. Aromachology and Its Application in the Textile Field. Fibres & Textiles in Eastern Europe 2005;13(6(54)) 41-44.

76. Buschmann HJ. Removal of Residual Surfactant Deposits from Textile Materials with the Aid of Cyclodextrins. Melliand Textilberchite. 1995(9).

77. Jellinnek JS. Aromachology: A Status Review. Perfume & Flavorist 1994;19 25-49.

78. Wu CH. Essential Oil and Aromatherapy. Kexue Nong Ye 1999;47(3) 1-3.

79. Mazzaro D. The home fragrance market. Chemical Market Reporter 2000;4 14.

80. Vraz Kresevic S, Voncina B, Vukusic Bischof S, Katovic D. Textile Materials Treated With Eco - Friendly Insecticide Agents. In: Dragcevic Zvonko (ed.): 4th International Textile, Clothing & Design Conference: Magic world of textiles: book of proceedings, ITC&DC, 05-08 October, 2008, Dubrovnik, Croatia. Zagreb: Faculty of Textile Technology, University of Zagreb; 2008.

81. Vraz Kresevic S, Voncina B, Gersak J. Insect resistant and eco-friendly textile materials. In: Dragcevic Zvonko (ed.): 4th International Textile, Clothing & Design Conference: Magic world of textiles : book of proceedings, ITC&DC, 05-08 October, 2008, Dubrovnik, Croatia. Zagreb: Faculty of Textile Technology, University of Zagreb; 2008.

82. Gao P, Nie X, Zou MJ, Shi YJ, Cheng G. Recent advances in materials for extended-release antibiotic delivery system. Journal of Antibiotics 2011;64(9) 625-634.

83. Blanchemain N, Karrout Y, Tabary N, Neut C, Bria M, Siepmann J, Hildebrand HF, Martel B. 2011Methyl-beta-cyclodextrin modified vascular prosthesis: Influence of the modification level on the drug delivery properties in different media. Acta Biomaterialia 2011;7(1) 304-314.

84. Wang JH, Cai ZS. A study of inclusion complex of miconazole nitrate with beta-cyclodextrin and its application on protein fabrics. In: Bai L,

Rui Y. (eds.) Researches and Progresses of Modern Technology on Silk, Textile And Mechanicals I: Conference publication of the 6th China International Silk Conference and the 2nd International Textile Forum, 13-14 September 2007, Suzhou, Jiangsu, P.R. China. Chemical Industry Press; 2007.

85. Salipira KL, Krause RW, Mamba BB, Malefetse TJ, Cele LM, Durbach SH. Cyclodextrin polyurethanes polymerised with multi-walled carbon nanotubes: Synthesis and characterisation. Materials Chemistry and Physics 2008;111(2-3) 218-224.

86. Salipira KL, Mamba BB, Krause RW, Malefetse TJ, Durbach SH. Cyclodextrin polyurethanes polymerised with carbon nanotubes for the removal of organic pollutants in water. Water SA 2008;34 113-118. http://www.wrc.org.za/downloads/watersa/2008/2198.pdf (accessed 3 July 2012).

87. Mamba BB, Krause RW, Malefetse TJ, Mhlanga SD, Sithole SP, Salipira KL, Nxumalo EN. Removal of geosmin and 2-methylisorboneol (2-MIB) in water from Zuikerbosch Water Treatment Plant (Rand Water) using β-cyclodextrin polyurethanes. Water SA 2007;33 223-228. http://www.wrc.org.za/downloads/watersa/2007/Apr%2007/2002.pdf (accessed 3 July 2012).

88. Prabaharan M, Mano JF. Chitosan derivatives bearing cyclodextrin cavities as novel adsorbent matrices. Carbohydrate Polymers 2006;63(2) 153-166.

89. Uyar T, Havelund R, Hacaloglu J, Besenbacher F, Kingshott P. Functional Electrospun Polystyrene Nanofibers Incorporating alpha-, beta-, and gamma-Cyclodextrins: Comparison of Molecular Filter Performance. ACS Nano 2010;4(9) 5121-5130.

Chapter 9

Degradation of textile dyes using immobilized lignin peroxidase-like metalloporphines under mild experimental conditions

Paolo Zucca, Antonio Rescigno, Manuela Pintus, Andrea C Rinaldi and Enrico Sanjust

ABSTRACT

Background

Synthetic dyes represent a broad and heterogeneous class of durable pollutants, that are released in large amounts by the textile industry. The ability of two immobilized metalloporphines (structurally emulating the ligninolytic peroxidases) to bleach six chosen dyes (alizarin red S, phenosafranine, xylenol orange, methylene blue, methyl green, and methyl orange) was compared to enzymatic catalysts. To achieve a green and sustainable process, very mild conditions were chosen.

Results

IPS/MnTSPP was the most promising biomimetic catalyst as it was able to effectively and quickly bleach all tested dyes. Biomimetic catalysis was fully characterized: maximum activity was centered at neutral pH, in the absence of any organic solvent, using hydrogen peroxide as the oxidant. The immobilized metalloporphine kept a large part of its activity during multi-cycle use; however, well-known redox mediators were not able to increase its catalytic activity. IPS/MnTSPP was also more promising for use in industrial applications than its enzymatic counterparts (lignin peroxidase, laccase, manganese peroxidase, and horseradish peroxidase).

Conclusions

On the whole, the conditions were very mild (standard pressure, room temperature and neutral pH, using no organic solvents, and the most environmental-friendly oxidant) and a significant bleaching and partial mineralization of the dyes was achieved in approximately 1 h. Therefore, the process was consistent with large-scale applications. The biomimetic catalyst also had more promising features than the enzymatic catalysts.

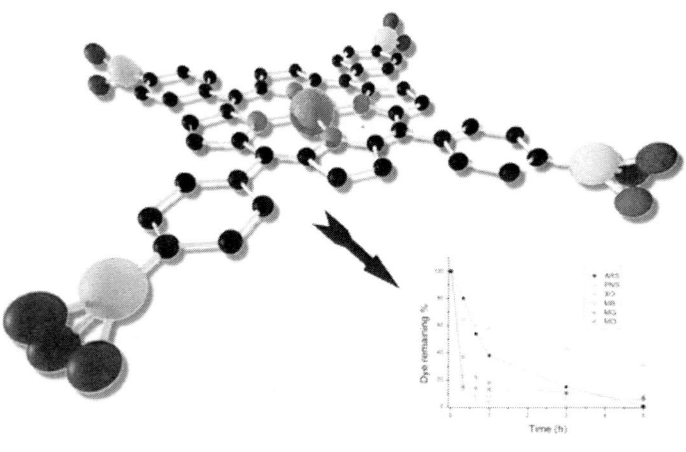

KEYWORDS

Biomimetic, Dye, Lignin-peroxidase, Porphine,

1. BACKGROUND

Textile dyes are a broad and heterogeneous class of molecules used in many technological fields (i.e., textile and paper production, food technology, and hair coloring) [1]. Many chemical classes of synthetic dyes are frequently employed on an industrial scale; azo dyes, triarylmethanes, phenothiazine, and anthraquinones are among the most widespread [2, 3].

Because the goal of the dyeing process is color durability, textile dyes of all chemical classes are commonly highly resistant to both chemical and physical degradation. Accordingly, textile dyes represent a serious environmental concern because of their chemical inertness, high annual production (over 10,000 tons per year [1]), toxicity/carcinogenicity [4, 5], and the large amount of dyes released in industrial wastewaters (up to 50% of total process intake [6]).

The European Directive 2002/61/EC forbids the use of some products (in particular some azo dyes). However, these restrictions only marginally solve the environmental problem because of the large volume of other dyes discharged every year.

Several physical, chemical and enzymatic approaches have been proposed to remove dyes from industrial wastewaters [7–16] that require long reaction times, extreme operational conditions, and high costs [1]. Accordingly, the development of a clean and environmentally friendly process for removing dyes is a challenging issue, far from being solved.

Redox-active metalloporphines are efficient catalysts; in the presence of a proper oxidant [17–21], they are able to oxidize many substrates, including several textile dyes, under very mild conditions. We previously studied the ability of an immobilized manganese porphine to oxidize alizarin red S (ARS), an anthraquinone dye) [22], and phenosafranine (PNS), a phenylphenazinium dye [23], elucidating several aspects of the catalytic mechanism.

In this study, we compare the catalysts IPS/MnTSPP and PP-PVA/FeTFPP (Figure 1), each containing a metalloporphine (Mn- and Fe-, respectively) immobilized onto a solid support, thus emulating ligninolytic peroxidases [24]. We investigated the ability of these two catalysts to oxidize different chemical classes of textile dyes in the presence of hydrogen peroxide under mild conditions. We also report the results of these catalysts on ARS and

PNS to allow a comprehensive comparison. All the dyes included in the study are reported in Table 1.

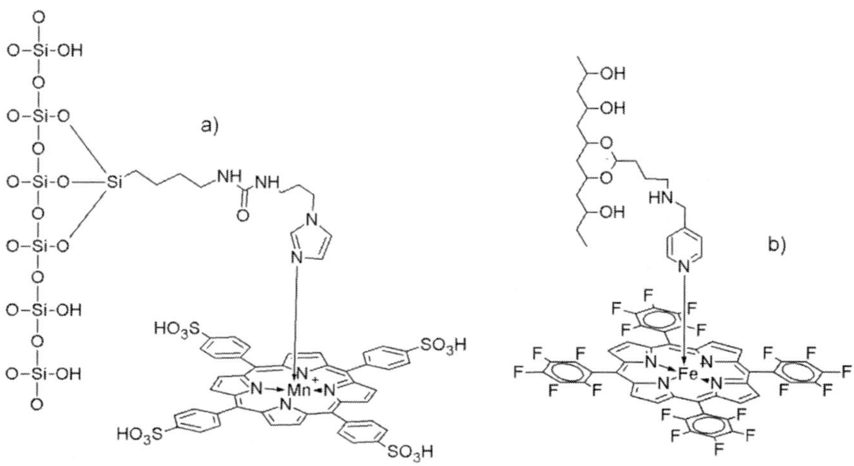

Figure 1. Proposed structure of the catalysts IPS/MnTSPP (a) and PP-PVA/FeTFPP (b). **For more details, see: [24, 25].**

2. RESULTS AND DISCUSSION

2.1 Comparison of IPS/MnTSPP and PP-PVA/FeTFPP

We previously developed two biomimetic catalysts, IPS/MnTSPP and PP-PVA/FeTFPP, which are able to oxidize several lignin-model compounds. IPS/MnTSPP strictly emulates a lignin peroxidase (LiP) active site, having an imidazole-grafted support that coordinates a Mn(III)-porphine [24]. This catalyst showed a LiP-like activity, whereas, unlike an enzyme such as manganese peroxidase (MnP), it was not able to use the Mn^{2+} ion as a redox mediator. Conversely, PP-PVA/FeTFPP was only “bioinspired” (i.e., only inspired by but not strictly emulating, a natural enzyme active site) because the metalloporphine was coordinated by a pyridine-grafted support [25] rather than imidazole, as in ligninolytic peroxidases. This catalyst showed both LiP-like and MnP-like catalytic activities.

Table 1. The six textile dyes included in this study

	Dye	MW (g/mol)	λ_{max} (nm)	ε (mol * L^{-1} * cm^{-1})
Alizarin Red S (ARS)	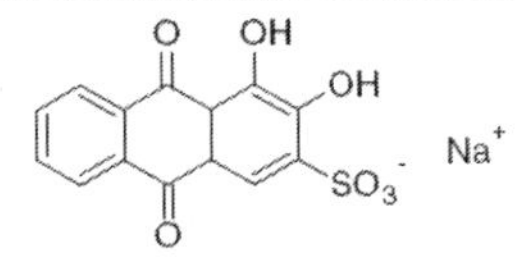	342.26	520	7200
Phenosafranine (PNS)	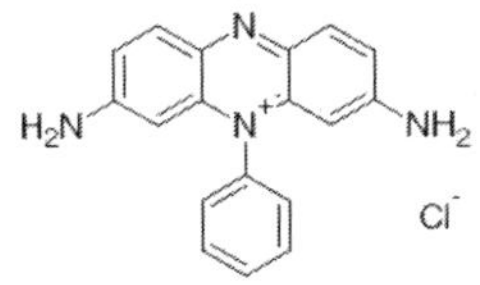	322.80	520	32000
Xylenol Orange (XO)	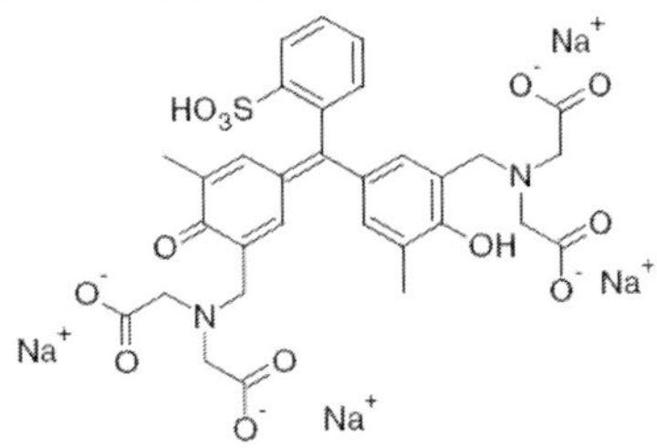	782.56	435	10500
Methylene Blue (MB)	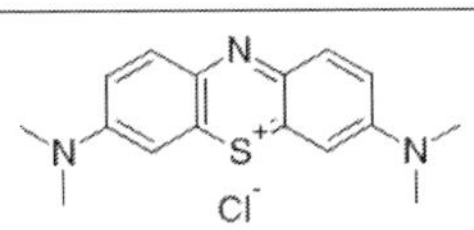	319.86	660	79500
Methyl Green (MG)	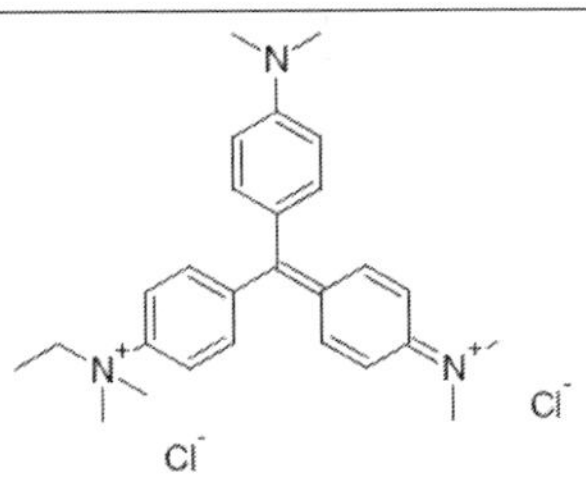	472.51	632	55700
Methyl Orange (MO)	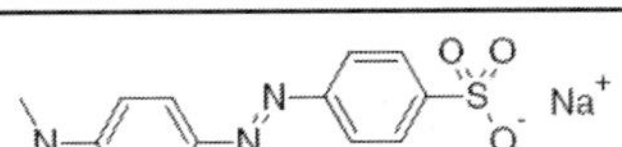	327.34	464	24800

These two catalysts were tested for their ability to decolorize several textile dyes in the presence and absence of Mn^{2+} using hydrogen peroxide as the oxidant in accordance with green chemistry principles [26]. The investigation employed very mild conditions (neutral pH and room temperature, using no organic solvents, and the most environmental-friendly oxidant) to serve as the basis for an environmentally sustainable textile wastewater treatment process.

The results of these tests are reported in Table 2. Both biomimetic catalysts easily degraded all tested dyes, showing wide substrate specificity. As previously reported [24], the catalytic behavior of IPS/MnTSPP was not affected by the presence of Mn(II), showing similar bleaching percentages in both the presence and absence of this ion. All six dyes were rapidly bleached under the conditions studied; more than with a removal in all cases higher than 50% of each dye (up to > 90% for PNS, MG, and MO) was removed in 1 h. In contrast, PP-PVA/FeTFPP was almost inactive in the absence of a suitable redox mediator. The conversion of all six dyes increased in the presence of Mn^{2+}, leading to values similar on average to those observed for IPS/MnTSPP.

Table 2. Removal of the studied dyes in the presence of hydrogen peroxide and either IPS/MnTSPP or PP-PVA/FeTFPP, in both the presence and absence of Mn(II) as the redox mediator

Mn(II)	Catalyst	Dyes removal % (1 h)					
		ARS	PNS	XO	MB	MG	MO
No	IPS/MnTSPP	62 ± 1	92 ± 2	42 ± 4	96 ± 4	82 ± 5	87 ± 2
	PP-PVA/FeTFPP	n.d.	n.d.	n.d.	n.d.	n.d.	n.d.
Yes	IPS/MnTSPP	55 ± 1	88 ± 2	44 ± 2	91 ± 3	77 ± 4	85 ± 4
	PP-PVA/FeTFPP	48 ± 3	58 ± 2	55 ± 5	37 ± 1	48 ± 2	91 ± 2

The reaction mixtures contained 5 mg of catalyst in 1 mL of 25 mM buffer solution containing 8.8 mM hydrogen peroxide and a proper concentration of the selected dyes (0.29 mM ARS, 0.31 mM PNS, 2 mM XO, 0.15 mM MB, 1.5 mM MG, and 1.25 mM MO). Quantification of the removed dye was obtained photometrically, using the λ_{max} and molar extinction coefficients reported in Table 1. n.d.: not detectable. ($n = 4$).

Considering that textile wastewaters do not usually contain manganous ion and that it would therefore be necessary to add Mn^{2+}, itself a heavy metal pollutant, to the reaction, IPS/MnTSPP was considered the most promising catalyst for future industrial applications Hence, the rest of study focused on IPS/MnTSPP.

The time dependence of dye bleaching by IPS/MnTSPP is shown in Figure 2. Note that most of the bleaching occurs in the first hour of the reaction, consistent with industrial timing needs.

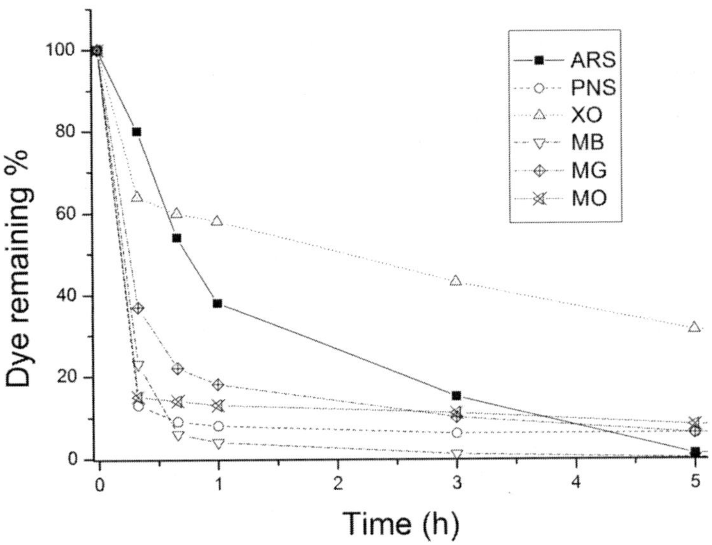

Figure 2. Time dependence of dye removal in the presence of IPS/MnTSPP and hydrogen peroxide. The reaction mixtures contained 5 mg of catalyst in 1 mL of 25 mM buffer solution containing 8.8 mM hydrogen peroxide and a proper concentration of the selected dyes (0.29 mM ARS, 0.31 mM PNS, 2 mM XO, 0.15 mM MB, 1.5 mM MG, and 1.25 mM MO). Quantification of the dyes was obtained photometrically, using the λ_{max} and molar extinction coefficients reported in Table 1.

2.2 Characterization of bleaching by IPS/MnTSPP and H_2O_2

IPS/MnTSPP catalysis was strongly pH-dependent; an approximately neutral pH was optimal for all dyes tested except XO and PNS, which had optimal bleaching at pH 6 and 8, respectively (data not shown). However, XO and

PNS were sufficiently bleached at pH 7 that IPS/MnTSPP remains a viable catalyst for environmentally friendly wastewater treatment.

An exhaustive kinetic modeling of this catalysis is beyond the scope of our work. However, the decolorization of all six dyes apparently followed saturation Michaelis-Menten-like kinetics, allowing the calculation of catalytic parameters (Table 3). Reported Michaelis constants (K_M) showed that IPS/MnTSPP had significantly higher affinity for cationic dyes than anionic dyes (lower K_M for PNS and MB, higher K_M for ARS and XO). This lower affinity led to the lowest bleaching rate for XO and ARS (compare Figure 2). Notably, in all cases, K_M values were in the millimolar range and did not suffer from substrate inhibition. This can be a crucial issue for potential industrial applications because some bleaching approaches (i.e., enzymatic, *vide ultra*) are not able to work at these concentrations.

To test the hypothesis of a one-electron abstraction mechanism as the first step of the bleaching process, the influence of several well-known redox mediators was also studied: 4-hydroxy-2,2,6,6-tetramethylpiperidine-1-oxyl (OH- TEMPO), *N*-hydroxyphthalimide (NPH), *N*-hydroxysuccinimide (NHS), and *N*-hydroxybenzotriazole (NHT). The results are summarized in Table 4. Only in the case of MG was an increase (three-fold) of the catalytic activity observed; in all the other cases, the redox-mediators mainly decreased IPS/MnTSPP catalysis (this quenching could be due to the lower reactivity of the oxygen-centered radical form of these mediators). Even in the best case, however, because the increase in catalytic activity was moderate, the industrial benefit of these redox mediators may not justify their environmental impact.

The catalyst was also investigated for its reuse potential. The results are reported in Table 5. On average, a rapid decrease of catalytic activity was observed in the first 2–3 cycles. Subsequently, activity was stable at approximately 25-30% of maximum for at least 8–10 cycles. A potential explanation for this pattern is the possibility that a fraction of some dyes remains tightly adsorbed on the catalyst even when the supernatant is completely decolorized. This adsorbed dye would likely reduce the catalytic performance by steric hindrance and/or electrostatic repulsion of incoming substrate molecules. Nevertheless, the catalyst keeps a significant part of its catalytic activity during multi-cycle employment. This is an important quality for large-scale applications.

Table 3. Kinetic parameters measured for the bleaching of each dye by IPS/MnTSPP and hydrogen peroxide

Sub-strate	Kinetic parameter	Dye					
		ARS	PNS	XO	MB	MG	MO
Reducing substrate	K_M (mM)	2.11 ± 0.74	0.32 ± 0.04	1.33 ± 0.38	0.11 ± 0.02	0.72 ± 0.12	0.63 ± 0.05
	k_{cat} (min^{-1})	0.14 ± 0.03	2.2 ± 0.1	98 ± 14	19.0 ± 1.8	108 ± 9	67 ± 4
	k_{cat}/K_M (mM^{-1} min^{-1})	0.066 ± 0.037	7.0 ± 1.2	74 ± 31	166 ± 52	150 ± 37	105 ± 16
H_2O_2	K_M (mM)	3.21 ± 0.12	3.0 ± 0.4	7.9 ± 1.9	2.4 ± 0.5	0.43 ± 0.08	7.4 ± 1.5
	k_{cat} (min^{-1})	0.13 ± 0.01	1.9 ± 0.1	78 ± 8.9	15.2 ± 1.3	159 ± 12	71 ± 8
	k_{cat}/K_M (mM^{-1} min^{-1})	0.040 ± 0.004	0.63 ± 0.13	9.9 ± 3.6	6.4 ± 2.1	369 ± 100	9.6 ± 3.2

Data for ARS and PNS were previously reported and are added here to allow a full comparison [22, 23] ($n = 5$).

Table 4. Effect of well-known redox-mediators on the bleaching of the six tested dyes by IPS/MnTSPP

Redox mediator	% Catalytic activity					
	ARS	PNS	XO	MB	MG	MO
OH-TEMPO	98	45	277	56	336	16
NHT	142	46	162	38	299	121
NHS	141	89	146	90	313	122
NPH	103	32	74	56	340	57

Results are expressed by referring to a 100% reference activity in the absence of any mediator. Data for ARS and PNS were previously reported and are added here to allow a full comparison [22, 23].

Table 5. Multi-cycle activity of IPS/MnTSPP

Catalytic cycle	% Catalytic activity					
	ARS	PNS	XO	MB	MG	MO
1	100	100	100	100	100	100
2	68	73	100	99	98	44
3	49	70	93	73	97	37
4	33	61	88	49	97	34
5	30	51	47	37	97	28
6	28	49	41	34	96	28
7	28	44	31	34	96	28
8	26	41	30	26	94	27
9	25	39	29	25	94	20
10	25	35	27	24	90	19

Catalytic activity of the IPS/MnTSPP adduct upon multicyclic use during the decolorization of the studied dyes ($n = 3$). Data for ARS and PNS were previously reported and are added here to allow a full comparison [22, 23].

Significant removal of total organic carbon (TOC) was also observed during the bleaching by H_2O_2/IPS/MnTSPP, as shown in Figure 3. In the case of ARS, PNS, and MB, over 50% of TOC removal was observed in 24 hours, whereas lower decreases were recorded for XO and MG. Conversely, in the case of MO no TOC modification occurred during the catalysis. These results suggest that a decarboxylative degradation occurred for some of the dyes, whose TOC values decreased significantly during the bleaching experiments.

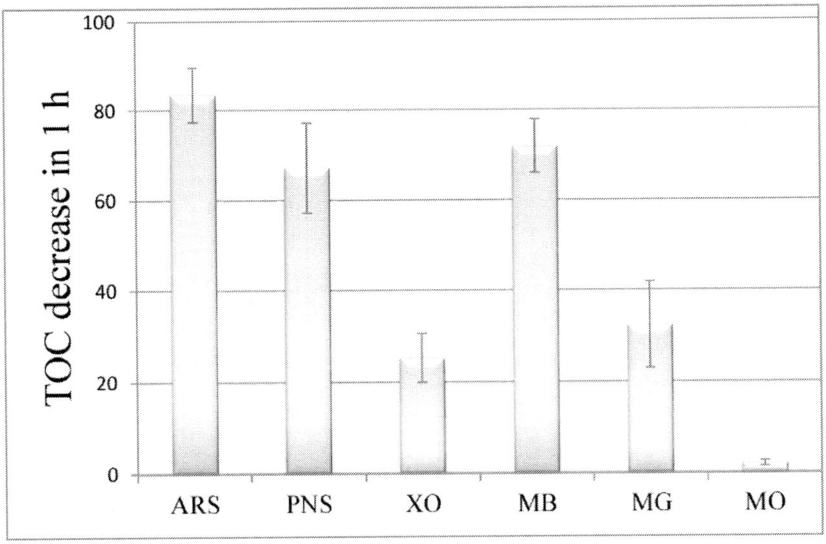

Figure 3. Total organic carbon (TOC) decrease during the bleaching of the textile dyes by IPS/MnTSPP (***n*** = 3).

IPS/MnTSPP was also tested for its ability to decolorize a complex mixture of all the six dyes, emulating realistic textile plant waste. The six dyes were each present in the saturating concentrations used above, leading to high impacting TOC (approximately 1580 mg/L). The changes in the UV–vis spectra (Figure 4) show a significant decolorization, although with a slower kinetics (approximately 5 hours for a 50% decrease of the main visible peak). However, the feasibility of using this treatment in an industrial process was confirmed by a significant TOC decrease (approximately 20%) in the same time frame.

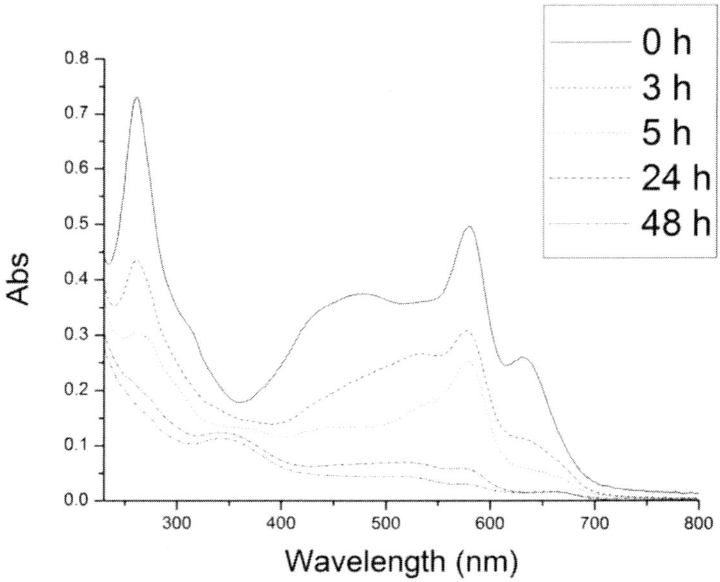

Figure 4. Changes in UV–vis spectra during the bleaching of a mix of the six studied dyes by IPS/MnTSPP.

2.3 Comparison with enzymatic bleaching

A full comparison with the enzymatic bleaching of the dyes was also performed. The ligninolytic enzymes tested were laccase (LC), lignin peroxidase, and the non-ligninolytic enzyme horseradish peroxidase (HRP). In addition, Mn(III) was also used as oxidizing agent, mimicking manganese peroxidase action [27]. These biological catalysts suffered from significant substrate inhibition, preventing their use at the dye concentrations used for IPS/MnTSPP. Usually, concentrations of dyes at least of one order of magnitude lower were necessary. These enzymes also suffer from H_2O_2 inhibition, needing lower concentrations of oxidizing substrate. Accordingly, under the same conditions, it is not surprising that enzymatic bleaching efficiency was significantly lower than IPS/MnTSPP bleaching efficiency (<1% dye removal in all cases). Enzymatic bleaching activity, under the conditions studied, is reported in Table 6, allowing a qualitative analysis of substrate specificity only.

Table 6. Enzymatic bleaching of the dyes

Dye	LC	LC + H_2O_2	HRP	LiP	Mn^{3+}
ARS	+	++	+/–	+/–	+
PNS	-	-	-	+/–	-
XO	+	+	+	-	+
MB	-	+/–	+/–	+/–	-
MG	+	+	++	+/–	++
MO	+	+/–	+	+/–	+

In the conditions tested, ligninolytic enzymes present a narrower substrate specificity towards the dyes included in this study than IPS/MnTSPP. Legend: LC laccase, HRP horseradish peroxidase, LiP lignin peroxidase, ++ good substrate (complete bleaching < 1 h), + substrate, +/– weak substrate (bleaching < 10% in 24 h), - no substrate.

None of the examined enzymes were able to bleach all the studied dyes. For instance, PNS is resistant to LC, HRP, and Mn^3+, whereas only a partial degradation with concomitant irreversible enzyme inactivation is observed with LiP. MB underwent only partial oxidation with the enzymatic catalysts.

IPS/MnTSPP is able to effectively bleach all the dyes studied, as described above. This feature, combined with its ability to oxidize relatively high concentration of dye (in the millimolar range), and with its partial multi-cycling, may make IPS/MnTSPP catalysis a feasible alternative for industrial treatment of textile wastewaters.

3. EXPERIMENTAL SECTION

3.1 Chemicals and instrumentation

All reagents were the best commercial grade available and were used without further purification. In particular, silica gel 100 came from Fluka (cat. No. 60746), and MnTSPP and FeTFPP were purchased from Sigma-Aldrich (cat. No. 441813 and 252913, respectively), PVA from Aldrich (cat. No. 363138), glutaraldehyde from Fluka (cat. no. 49629), horseradish peroxidase (E.C. 1.11.1.7) and lignin peroxidase (E.C. 1.11.1.14) were from Sigma-Aldrich (cat No. P-6782 and 42603, respectively). Laccase from*Pleurotus sajor-caju* was purified through two chromatographic steps (details in [28, 29]).

Spectrophotometric measurements were carried out with an UltroSpec 2100pro (Amersham Bioscience). Concentrations of the substrates were also measured by UV-HPLC and GC-MS, using previously described methods [30, 31].

3.2 Preparation of IPS/MnTSPP

The complex IPS/MnTSPP was synthesized as previously described [24]. Briefly, 10 g plain silica gel was functionalized by overnight reaction at 80°C with 10 mmol 3-(1-imidazolyl)propylcarbamoyl-3'-aminopropyl-triethoxysilane solubilized in 10 mL dioxane, in the presence of 10 μL *N*-methylmorpholine acting as the catalyst. The activated silica (IPS) was washed consecutively with 0.5 M HCl, H_2O, 0.1 M NaOH and H_2O again. The wet silica was then carefully dried overnight in a vacuum oven.

The bound MnTSPP was quantified by the difference in absorption at 400 nm (the λ_{max} of the MnTSPP Soret band), as described in [24].

3.3 Preparation of PP-PVA/FeTFPP

The complex PP-PVA/FeTFPP was synthesized as previously described [25]. Briefly, 1 g of aminopropyl crosslinked PVA (prepared as in [32]) was suspended in an excess of 0.05 M sodium acetate buffer pH 5, and treated with 0.1 mL 4-pyridinecarboxaldehyde and 0.5 g sodium cyanoborohydride. After 24 h of incubation at 25°C, the support was exhaustively washed with water, and dried at 50°C, to obtain PP-PVA. One-gram aliquots of PP-PVA were treated with 20 mg of FeTFPP, solubilized in 10 mL DMSO. The slurries were kept stirring in the dark for 24 h, washed exhaustively with DMSO, and carefully dried at 50°C.

Bound metalloporphine was quantified by difference through spectrophotometric measurement at 411 nm (ε_{411} = 115,000 M^{-1} cm^{-1} in DMSO) [25].

3.4 Activity measurements

Routine measurements of the catalytic activity of immobilized metalloporphines were performed through spectrophotometric assays. The suspension contained 5 mg of catalyst in 1 mL of 25 mM buffer solution containing 8.8 mM hydrogen peroxide and a proper concentration of the selected dye. The routine analyses were performed using the following concentrations of dyes (chosen to be only slightly higher than respective K_M, to better observe influences on catalytic activity): 0.29 mM ARS, 0.31 mM PNS, 2 mM XO, 0.15 mM MB, 1.5 mM MG, and 1.25 mM MO. During the kinetic experiments, wide ranges of substrate concentrations were studied (0.01-2 mM for the dyes, 0.1-40 mM for hydrogen peroxide).

The reaction mixture was then stirred at 25°C in the dark, and the reaction was monitored photometrically using the λ_{max} and molar extinction coefficients, reported in Table 1.

The actual hydrogen peroxide concentration was determined by $KMnO_4$ titration.

When the reaction pH was not 7, the pH was corrected to 7 with 0.1 mL of potassium phosphate buffer 1 M (pH 7) before an absorbance measurement. UV–vis spectra in the range 230–700 nm were also recorded. Different McIlvaine buffers were used in the pH range 3–8.

When used, redox mediators (4-hydroxy-2,2,6,6-tetramethylpiperidine-1-oxyl OH- TEMPO; *N*-hydroxyphthalimide NPH; *N*-hydroxysuccinimide NHS; and *N*-hydroxybenzotriazole NHT) were added to the reaction medium at a final concentration of 1 mM. For the MnP-like assays, all the experiments were repeated as above in the presence of 1 mM $MnSO_4$ and 50 mM malonic acid. During multi-cyclic runs, the catalysts were regenerated between consecutive cycles by repeated washings with water and 2-propanol and by drying at 50°C.

Total organic carbon determination was performed by automatic analyzer TOC 5000A (Shimadzu), equipped with Pt-Al_2O_3 oxidizing catalyst. The volume of injection was 25 μL. The calibration curve was obtained using a range of standard solutions of potassium phthalate (0–200 ppm).

Michaelis-Menten kinetic parameters were calculated by using R 2.5.1 software (*R Foundation for Statistical Computing*, Vienna).

3.5 Enzymatic assays

When horseradish peroxidase (HRP) was used, up to 20 E.U. were present in a final volume of 1 mL of 25 mM buffer (pH 4, 5, 6, or 7) with 8.8 mM H_2O_2. Due to significant substrate inhibition, lower dyes concentrations were used: 119 μM ARS; 31 μM PNS; 200 μM XO; 15 μM MB; 150 μM MG; 65 μM MO.

The same conditions were adopted for the experiments with laccase (23 E.U. added); every experiment was performed in the presence and absence of H_2O_2.

In the case of lignin peroxidase (LiP), 0.2 E.U. were present in the final volume of 1 mL, with 0.176 mM H_2O_2, due to strong substrate inhibition.

Moreover, to mimic manganese peroxidase (MnP) action, other experiments were conducted using Mn(III) as the putative oxidizing agent. One mM manganese triacetate was dissolved in 50 mM sodium malonate buffer, pH 6.5 [27], and the final mixture contained this solution along with dyes at the same concentrations, as mentioned above.

4. CONCLUSION

IPS/MnTSPP has proved able to efficiently degrade all tested dyes, regardless of their chemical class, under very mild conditions: standard pressure, room temperature and neutral pH, using no organic solvents, and the most environmental-friendly oxidant. The yields and time requirements of the process were quite promising for a potential industrial application. Furthermore, a multi-cycle capability was described, even if only 30% of catalytic activity is retained for 8–10 cycles.

In comparison to enzymatic systems, the biomimetic catalyst was more efficient and had wider substrate specificity. Therefore, after proper optimization of the process, this catalyst is worth consideration as a possible medium/large scale application in textile wastewater treatment.

ACKNOWLEDGMENTS

This research was partly funded by Fondazione Banco di Sardegna (Project Code 861/2012) and Regione Autonoma Sardegna (PO Sardegna FSE 2007–2013, L.R.7/2007, Bando Giovani Ricercatori, Project CRP1_27, and PO Sardegna FSE 2007–2013, L.R.7/2007, Ricerca di Base Bando 2008, Project CRP1_408).

AUTHORS' CONTRIBUTIONS

PZ performed the main part of the experiments and drafted the manuscript. MP performed TOC and enzymatic assays. ACR, and AR helped analyzing the data and draft the manuscript. ES conceived the study, coordinated the study, and worked on data analysis and interpretation. All the authors read and approved the final manuscript.

REFERENCES

1. Forgacs E, Cserhati T, Oros G: Removal of synthetic dyes from wastewaters: a review. Environ Int. 2004, 30: 953-971.
2. Azizi A, Moghaddam MRA, Arami M: Application of wood waste for removal of reactive blue 19 from aqueous solutions: optimization through response surface methodology. Environ Eng Manag J. 2012, 11: 795-804.
3. Brînzilă CI, Ciobanu R, Brillas E: Effect of experimental parameters on crystal violet mineralization by electro-fenton processes. Environ Eng Manag J. 2012, 11: 517-523.
4. Tsuda S, Murakami M, Matsusaka N, Kano K, Taniguchi K, Sasaki YF: DNA Damage induced by red food dyes orally administered to pregnant and male mice. Toxicol Sci. 2001, 61: 92-99.
5. Walthall WK, Stark JD: The acute and chronic toxicity of two xanthene dyes, fluorescein sodium salt and phloxine B, to Daphnia pulex. Environ Pollut. 1999, 104: 207-215.

6. Pierce J: Colour in textile effluents - the origins of the problem. J Soc Dye Colour. 1994, 110: 131-133.

7. Colar LA, Ilinoiu EC, Manea F, Pode R, Lazau C: Photocatalytic activity of a silver doped TIO 2 modified zeolite in the degradation of reactive yellow 125 azo dye. Environ Eng Manag J. 2012, 11: 61-68.

8. Zucca P, Rescigno A, Olianas A, Maccioni S, Sollai F, Sanjust E: Induction, purification, and characterization of a laccase isozyme from Pleurotus sajor-caju and the potential in decolorization of textile dyes. J Mol Catal B Enzym. 2011, 68: 216-222.

9. Pakshirajan K, Jaiswal S, Das RK: Biodecolourization of azo dyes using phanerochaete chrysosporium: effect of culture conditions and enzyme activities. J Sci Ind Res. 2011, 70: 987-991.

10. Córdoba A, Magario I, Ferreira ML: Experimental design and MM2-PM6 molecular modelling of hematin as a peroxidase-like catalyst in Alizarin Red S degradation. J Mol Catal A Chem. 2012, 355: 44-60.

11. Zhu L, Ghosh T, Park CY, Meng ZD, Oh WC: Enhanced sonocatalytic degradation of rhodamine B by graphene-TiO 2 composites synthesized by an ultrasonic-assisted method. Chin J Catal. 2012, 33: 1276-1283.

12. Oh WC, Chen M, Cho K, Kim C, Meng Z, Zhu L: Synthesis of graphene-CdSe composite by a simple hydrothermal method and its photocatalytic degradation of organic dyes. Chin J Catal. 2011, 32: 1577-1583.

13. Asgher M, Kamal S, Iqbal HMN: Improvement of catalytic efficiency, thermo-stability and Dye decolorization capability of pleurotus ostreatus IBL-02 laccase by hydrophobic Sol Gel entrapment. Chem Centr J. 2012, 110:

14. Stengl V, Matys Grygar T, Henych J, Kormunda M: Hydrogen peroxide route to Sn-doped titania photocatalysts. Chem Centr J. 2012, 113:

15. Velmurugan R, Sreedhar B, Swaminathan M: Nanostructured AgBr loaded TiO 2: an efficient sunlight active photocatalyst for degradation of reactive Red 120. Chem Centr J. 2011, 5:

16. Chiavola A: Textiles. Water Environ Res. 2012, 84: 1511-1532.

17. Zucca P, Rescigno A, Sanjust E: Ligninolytic peroxidase-like activity of a synthetic metalloporphine immobilized onto mercapto-grafted crosslinked PVA inspired by the active site of cytochrome P450. Chin J Catal. 2011, 32: 1663-1666.

18. Faria AL, Mac Leod TOC, Barros VP, Assis MD: Hydrocarbon oxidation catalyzed by iron and manganese porphyrins anchored on aminofunctionalized supports. J Brazil Chem Soc. 2009, 20: 895-906.

19. Dos Santos JS, Faria AL, Da Silva Amorin PM, La Luna FM, Caiado KL, Silva DOC E, Sartoratto PPC, Assis MD: Iron(III) porphyrin covalently supported onto magnetic amino-functionalized nanospheres as catalyst for hydrocarbon and herbicide oxidations. J Brazil Chem Soc. 2012, 23: 1411-1420.

20. Díaz-Díaz G, Blanco-López MC, Lobo-Castañón MJ, Miranda-Ordieres AJ, Tuñón-Blanco P: Hemo-acrylic polymers as catalyst for the oxidative dehalogenation of 2,4,6-trichlorophenol. Chloroperoxidase's mimic imprinting effects. J Mol Catal A Chem. 2012, 353–354: 117-121.

21. Tian P, Gao B, Chen Y: Synchronous synthesis and immobilization of pyridyl porphyrins on crosslinked polystyrene microspheres and catalytic oxidation performance of immobilized cationic Co-porphyrins. Chin J Catal. 2011, 32: 483-489.

22. Zucca P, Vinci C, Sollai F, Rescigno A, Sanjust E: Degradation of Alizarin Red S under mild experimental conditions by immobilized 5,10,15,20-tetrakis(4-sulfonatophenyl)porphine-Mn(III) as a biomimetic peroxidase-like catalyst. J Mol Catal A Chem. 2008, 288: 97-102.

23. Zucca P, Vinci C, Rescigno A, Dumitriu E, Sanjust E: Is the bleaching of phenosafranine by hydrogen peroxide oxidation catalyzed by silica-supported 5,10,15,20-tetrakis-(sulfonatophenyl)porphine-Mn(III) really biomimetic?. J Mol Catal A Chem. 2010, 321: 27-33.

24. Zucca P, Mocci G, Rescigno A, Sanjust E: 5,10,15,20-Tetrakis(4-sulfonato-phenyl)porphine-Mn(III) immobilized on imidazole activacted silica as a novel lignin-peroxidase-like biomimetic catalyst. J Mol Catal A Chem. 2007, 278: 220-227.

25. Zucca P, Sollai F, Garau A, Rescigno A, Sanjust E: Fe(III)-5,10,15,20-Tetrakis(pentafluorophenyl)porphine supported on pyridyl-

functionalized, crosslinked poly(vinylalcohol) as a biomimetic versatile-peroxidase-like catalyst. J Mol Catal A Chem. 2009, 306: 89-96.

26. Cojocariu AM, Mutin PH, Dumitriu E, Vioux A, Fajula F, Hulea V: Removal of dimethylsulfoxide from wastewater using mild oxidation with H2O2 over Ti-based catalysts. Chemosphere. 2009, 77: 1065-1068.

27. Wariishi H, Valli K, Gold MH: Manganese(II) oxidation by manganese peroxidase from the basidiomycete phanerochaete-chrysosporium - kinetic mechanism and role of chelators. J Biol Chem. 1992, 267: 23688-23695.

28. Rescigno A, Zucca P, Flurkey A, Inlow J, Flurkey WH: Identification and discrimination between some contaminant enzyme activities in commercial preparations of mushroom tyrosinase. Enzyme Microb Technol. 2007, 41: 620-627.

29. Sollai F, Zucca P, Sanjust E, Steri D, Rescigno A: Umbelliferone and esculetin: inhibitors or substrates for polyphenol oxidases?. Biol Pharm Bull. 2008, 31: 2187-2193.

30. Zucca P, Littarru M, Rescigno A, Sanjust E: Cofactor recycling for selective enzymatic biotrasformation of cinnamaldehyde to cinnamyl alcohol. Biosci Biotechnol Biochem. 2009, 73: 1224-1226.

31. Rescigno A, Casañola-Martin GM, Sanjust E, Zucca P, Marrero-Ponce Y: Vanilloid derivatives as tyrosinase inhibitors driven by virtual screening-based QSAR models. Drug Test Anal. 2011, 3: 176-181.

32. Sanjust E, Cocco D, Curreli N, Rescigno A, Sollai F, Bannister JV: Flavin-grafted poly(vinyl alcohol): preparation and properties. J Appl Polym Sci. 2002, 85: 2471-2477.

Chapter 10

Antibacterial Zinc Oxide Nanoparticle Coating of Polyester Fabrics

Claudia Rode[2*], Michael Zieger[1], Ralf Wyrwa[2], Susanne Thein[2], Cornelia Wiegand[1], Monika Weiser[3], Andreas Ludwig[4], Dirk Wehner[4], Uta-Christina Hipler[1]

[1]Department of Dermatology, Jena University Hospital, Jena, Germany

[2]INNOVENT e.V., Biomaterials Department, Jena, Germany

[3]Textile Research Institute Thuringia-Vogtland, Greiz, Germany

[4]Thorey Gera Textilveredelung GmbH, Gera, Germany

ABSTRACT

Polyester knitted fabrics have been equipped with antibacterial properties by coating with aque- ous suspensions of zinc oxide (ZnO) with different particle diameters. It can be shown that the antibacterial efficacy against Staphylococcus aureus and Klebsiella pneumonia and the permanence of the coating after defined washing cycles depends on the composition of the ZnO dispersion. To improve the wettability of the textile, the combustion chemical vapor deposition (CCVD) is used to generate a hydrophilic nano layer of a hydroxyl terminated silicon network on the surface of the textiles. The comparison of CCVD-treated and untreated polyester samples does not show differences in the antibacterial activity. The influence of the coating in terms of amount and particle size of ZnO on the viability and proliferation of 3T3 mouse fibroblast cells is examined. A particle content of a maximum of 20 μg ZnO/cm^2 is found to be cytocompatible for coating of textiles.

KEYWORDS

Antibacterial, Textile, ZnO Nanoparticles, Cytocompatibility

1. INTRODUCTION

Omnipresent colonization of material surfaces, especially by bacteria, is an undesired occurrence in the technical sector as well as in everyday life and can be life-threatening in the clinical area. In the case of used clothing and other fabrics, bacterial colonization causes unpleasant odors or infections. In recent years, mainly technical textiles are equipped with antibacterial properties, but nowadays the growing awareness of health and hygiene increases the demand for antibacterial materials in a wide range of applications including medicine, medical care, protective clothing, and household as well as sport and leisure time. The use of several substances for reducing bacterial colonization on textile surfaces have been reported including positively charged molecules [1] - [4] , silver [4] - [7] , copper [8] , triazin derivatives [9] [10] , chitosan [3] [11] [12] , or triclosan [3] . Because of existing disadvantages like toxicity [13] , high costs or issues of bacterial resistance of the mentioned examples [14] - [16] , we decide to apply zinc oxide (ZnO) as antimicrobial agent [17] - [19] . In several studies the antimicrobial effect of zinc oxide coatings against different bacteria can be already verified [20] - [23].

For the development of antibacterial textiles for applications on a large scale, e.g. in medicine and for wearing apparel, it is necessary to find the optimum of ZnO concentration to effectively combat bacteria while preventing cytotoxic effects. Therefore, the main focus of our study is the investigation of the correlation between the ZnO content, the antibacterial activity and the cytocompatibility of ZnO coated textiles. Furthermore, our research addresses the improvement of the coated polyester regarding the optical appearance and especially of the permanence of the antibacterial effect. For that reason the CCVD process is applied to enhance the adhesive strength between the coating and the textile surface. The antibacterial activity of ZnO nanoparticle coated textiles has been proven using the nosocomial pathogens Staphylococcus aureus and Klebsiella pneumoniae.

2. MATERIALS AND METHODS

2.1. Materials

Powder of ZnO nanoparticles was obtained from ABCR (Karlsruhe, Germany, article no. AB 255364) and IBUtec advanced materials AG (Weimar , Germany). In addition, ZnO nanoparticle dispersion in water (40% (w/w)) was obtained from IBUtec.

Impranil DLU (aliphatic polycarbonate ester polyether polyurethane dispersion in water, 60% solid content), Baybond PU 406 (non-ionic polyurethane dispersion in water, 34.5% solid content) and Desmodur DN (aliphatic poly-isocyanate) were provided by Bayer Material Science AG (Leverkusen, Germany).

3T3 mouse embryonic fibroblasts were purchased from LGC Standards GmbH (Wesel, Germany).

Staphylococcus aureus ATCC 6538 and Klebsiella pneumoniae ATCC 4352 were obtained from the Leibniz Institute DSMZ-German Collection of Microorganisms and Cell Cultures (Braunschweig, Germany). For cultivation of bacteria, special peptone and "lab-lemco" powder for preparation of Caso bouillon and bacteriological agar were purchased from Oxoid Deutschland GmbH (Wesel, Germany). Columbia agar plates with 5% sheep blood were acquired from bioMérieux Deutschland GmbH (Nürtingen, Germany). Physiological NaCl solution (0.9%) was purchased from Fresenius Kabi Deutschland AG (Bad Homburg, Germany). Tween 20 was obtained from Carl Roth GmbH (Karlsruhe, Germany).

Polyester knitted fabrics were provided by Thorey Gera Textilveredelung GmbH (Gera, Germany).

2.2. Coating Process

Activation of the polyester was performed using the CCVD technique with a custom-made winding construction with two of four burners and a limited coating width of 30 cm. For the process, an air flow of 600 L/min was used. The air/gas ratio was adjusted to 18.8. The precursor (30% hexamethyldisiloxane (HMDSO) in isopropanol) was added at a flow rate of 5.5 mL/min. The speed of sweep was 50 m/min and the temperature of the guide rollers was kept at 50°C.

For preparation of the coating suspension, the ZnO powder (ABCR) or the 40% (w/w) ZnO suspension (IBUtec) were diluted with water to obtain the desired concentration of 0.25% to 1% of ZnO. ABCR nanoparticles were used as received without disaggregating. To the suspension 2% to 5% (w/w) of a binder (Baybond PU and Impranil DLU, respectively) was added.

For laboratory scale a culture dish was used for dip coating of the textiles. After storing the textiles for 5 min in the coating solution, they were removed and heated to 80°C for 15 min and then heated to 150°C for 30 min. During the coating procedure stirring at 750 rpm was necessary to avoid sedimentation of the nanoparticles.

The antibacterial coating in industrial scale was realized by a foulard machine (model HVF, Mathis AG, Oberhasli, Switzerland). Stirring was not possible during the coating procedure. The polyester was run with 100 cm/min through the ZnO suspension and squeezed at 0.3 MPa. Afterwards it was dried for 2 min at 100°C and the crosslinking reaction was performed by heating in a second step for 1.5 min at 150°C (Labdryer, Mathis AG, Oberhasli, Switzerland).

2.3. Characterization

Scanning electron microscopy (Supra 55V, Carl Zeiss SMT, Oberkochen, Germany) was used to investigate the size of the nanoparticles and the surface of coated polyester fabrics. Qualitative SEM-EDX analysis was carried out using the Quantax system with a Si (Li) detector (Röntec, Berlin, Germany). The samples were vapor coated with carbon and activated by an electron beam of 15 keV.

The zinc content of coated samples was determined by complexometric titration with a 0.1 M aqueous solution of ethylenediaminetetraacetic acid disodium salt (Aldrich, Steinheim, Germany). The zinc oxide on sample surface was dissolved by the use of 18% hydrochloric acid before analysis.

2.4. Washing Process

To evaluate the quality of antimicrobially finished fabrics, the permanence against mechanical and chemical st- ress was studied applying washing and drying cycles. The minimum number of cycles was defined correspond- ing to the designated purpose. The investigation was performed using an aqueous detergent solution as well as subsequent rinsing and drying steps in a modified wash test according to DIN EN ISO 105-C10, method A.

The washing tester "Linitest" was used for determining the wash durability of the fabrics. Two specimens with a size of 10 cm × 4 cm were treated in a steel container for 30 min at 40°C using a liquor ratio (specimen: liquor) of 50 g/L. The used washing liquor contained 5 g/L of a detergent.

The samples were transferred to a container with 2 L of distilled water and rinsed by careful moving. The samples were then rinsed under running cold water and dried in the air.

2.5. Antibacterial Activity

Testing for antibacterial activity was carried out in accordance to the Japanese Industrial Standard (JIS L 1902:2002, "Testing method for antibacterial activity of textiles") as reported previously [24] . In brief, S. aureus or K. pneumoniae were cultivated in caso-bouillon at 37°C for 18 h under aerobic conditions. For experiments, bacteria cultures were diluted with Caso bouillon to an inoculum concentration of 5×10^5 cfu/mL. 400 mg samples of the test material were inoculated with 200 μL test microbe inoculum and incubated at 37°C for 24 h under aerobic conditions. Untreated polyester was used as negative control. For determination of the germ number after 24 h, samples were extracted in 0.9% NaCl solution supplemented with 0.2% tween 20. Serial dilutions were plated onto Columbia agar plates and incubated for 24 h at 37°C. Afterward, colonies were calculated, total cfu (colony forming units) estimated, and growth reduction assessed according to Equation (1). According to JIS L 1902, a logarithmic microbial growth reduction of less than 0.5 denotes no antibacterial activity. Values from 0.5 to 1 are estimated as a slight, values greater than 1 and less or equal to 3 as a significant, and a log reduction greater than 3 as a strong antibacterial activity.

$$\log \text{growth reduction}_{(24h)} = \log \text{cfu}(\text{negative control})_{(24h)} - \log \text{cfu}(\text{sample})_{(24h)} \quad (1)$$

Equation (1) is the calculation of logarithmic microbial growth reduction.

2.6. Cytocompatibility

To assess cytocompatibility, the coated textiles and an uncoated sample as control were treated with 70% ethanol, washed with phosphate buffered saline (PBS) and seeded with 3T3 mouse fibroblast cells at a cell density of about 25,000 cells/cm^2. The cells were cultured in Dulbecco's Modified Eagle Medium (DMEM) containing 10% fetal bovine serum (FBS), 50 U/mL penicillin and 50 µg/mL streptomycin at 37°C under a 5% CO_2 atmosphere. The medium was renewed on day 2. After 1 and 4 days the medium was changed against fluorescein diacetate (FDA) and GelRed®. An Axiotech microscope (Zeiss, Jena, Germany) and a halogen lamp were used to monitor green and red fluorescence. Photomicrographs were recorded using a CCD microscope imaging system MP 5000 (Intas, Goettingen, Germany). Imaging was supported by Image-Pro Plus 5.1 software (Media Cybernetics, Silver Spring, USA). The total number of cells and the percentage of living cells were calculated after counting green fluorescent living cells and red fluorescent nuclei of dead cells.

3. RESULTS AND DISCUSSION

3.1. Nanoparticles

The size of ZnO nanoparticles influences color, release of Zn ions, and particle distribution as well as roughness of the coated surface. Therefore, ZnO nanoparticles which are differ in particle size should be used to achieve optimal properties of an antibacterial coating. SEM-images of the applied nanoparticles of ABCR as well as IBUtec are shown in Figure 1.

It was found that the particles of the two suppliers vary distinctly in particle size. The mainly hexagonal prismatic ZnO NPs obtained from ABCR show a white color and a particle size in the range of 50 - 300 nm with scattered larger particles and aggregates. The ZnO particles obtained from IBUtec have a spherical shape with an average particle size of 10 nm. Due to the small size these particles exhibit a brownish color.

3.2. Coating Process

For laboratory scale, polyester samples of 4 cm × 10 cm were dip coated in a culture dish using 100 mL of an aqueous ZnO suspension (0.3% - 2.0% (w/w)) containing Impranil DLU (2% (w/w)) as binder. The binder serves as a matrix for incorporating the ZnO nanoparticles improving the adhesion of ZnO to the hydrophobic textile surface. Complexometric analyses were performed at coatings with a concentration of 1.4% (w/w) ZnO. An average amount of 0.23 mg ZnO/cm^2 was found on these samples measured. Neither the particles used (ABCR or IBUtec) nor the activation with CCVD had an influence on the ZnO content of coated samples.

For coating procedures on industrial scale with a foulard machine a ZnO containing suspension (0.04% - 0.65% (w/w)) was used. Figure 2 shows graphically the determined amounts of ZnO of coated polyester samples by complexometric titration.

No linear correlation was found between applied amount of ZnO powder used from ABCR and measured amount of ZnO of coated textiles (Figure 2(a)). The coating solution with 0.32% ZnO content resulted in higher ZnO depositions on textile with CCVD pretreatment than the solution with 0.65% ZnO. This is most likely due to the deposition of the powder at the vessel bottom where it is not available during the coating procedure. This problem was avoided using the finely dispersed aqueous solution of ZnO nanoparticles from IBUtec (Figure 2(b)). Here, the measured amount of coating solution containing 0.5% of ZnO was found to be higher than the amount by using a 0.25% ZnO solution. With CCVD an even distinctly higher quantity of ZnO could be achieved because of the higher absorption properties of the hydrophilic nano layer on the surface. However, due to the incompatibility of the ZnO nanoparticle suspension with the binder

Impranil, the system had to be changed to use Baybond PU for coating. Additional use of the hardener Desmodur DN leads to further increase of the ZnO content on the textiles.

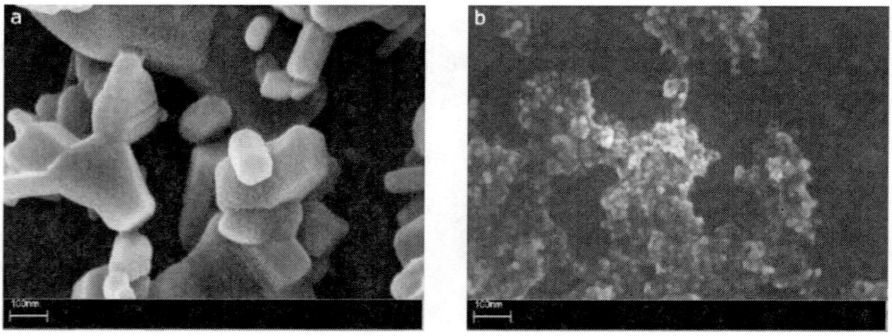

Figure 1. SEM images of ZnO nanoparticles obtained from ABCR (a) and IBUtec (b).

The SEM images of laboratory coated textiles reveal the considerably higher adhesion of the ZnO/Impranil system in the case of ABCR particles (Figure 3(a)) compared to the IBUtec particles (Figure 3(b)). Nevertheless, the qualitative EDX spectra (Figure 3(c), Figure 3(d)) show the presence of Zn in both coating systems. But in the case of ABCR ZnO particles palpable white deposits on the textiles could be observed (Figure 3(e)). However the appearance of the coating clearly favors IBUtec particles despite the slight brownish color (Figure 3(f)) whereby spotting can be avoided by uniform application at industrial conditions with the foulard machine.

3.3. Antibacterial Activity

Figure 4 shows the antibacterial activity of laboratory coated polyester samples against Staphylococcus aureus and Klebsiella pneumoniae in vitro. Coating with ABCR ZnO particles and Impranil DLU as binder conveyed a strong antibacterial activity against S. aureus and K. pneumoniae up to 10 washing cycles (with and without CCVD pretreatment). In contrast, coating with IBUtec ZnO particles (with and without CCVD pretreatment) was less permanent and accomplished only a significant antibacterial effect against

S. aureus and a slight antibacterial influence on K. pneumoniae after 10 washing cycles. After 50 washing cycles, textiles coated with ZnO nanoparticles of ABCR still obtained significant to strong results whereas coatings with ZnO particles of IBUtec showed a distinctly diminished antibacterial efficacy, though other studies showed an increasing activity with decreasing the particle size [22] [25] . We propose that the differences observed in the antibacterial activity of the different particles after several numbers of washing cycles are most likely due to a higher durability of ABCR ZnO particles on the polyester samples. Certainly the result based on the larger surface area of the IBUtec nanoparticles. These smaller particles will be dissolved faster during washing process.

Figure 5 shows the test results for the antimicrobial effect of the coated and washed polyester samples with the system ZnO (IBUtec, aqueous dispersion) and Baybond PU against Staphylococcus aureus and Klebsiella pneumoniae on industrial scale. Because of sedimentation of ABCR particles the IBUtec dispersion was used only for further studies. Simultaneously, the influences of the CCVD pretreatment and an additionally cross linking hardener (Desmodur DN) were investigated. No additional beneficial effect of the cross linking hardener on the washing stability of the coating could be observed. Pretreatment of the polyester samples with CCVD also showed no improvement of the permanence of the antibacterial activity. However, since pretreated textiles are more hydrophilic they can absorb more solution and the resulting amount of zinc oxide is higher than on untreated textiles. Figure 5 also shows that the washing stability by using Baybond PU is not as good as using Impranil DLU. Mainly slight antimicrobial effects could be observed after 10 washing cycles thus the influence of more than 10 washing cycles to the antimicrobial activity was not determined.

3.4. Cytocompatibility

The cytocompatibility of the coated polyester samples was studied by seeding 3T3 fibroblast cells on the surface of the textiles and then estimating cell adhesion and viability using a FDA/GelRed®live/dead assay.

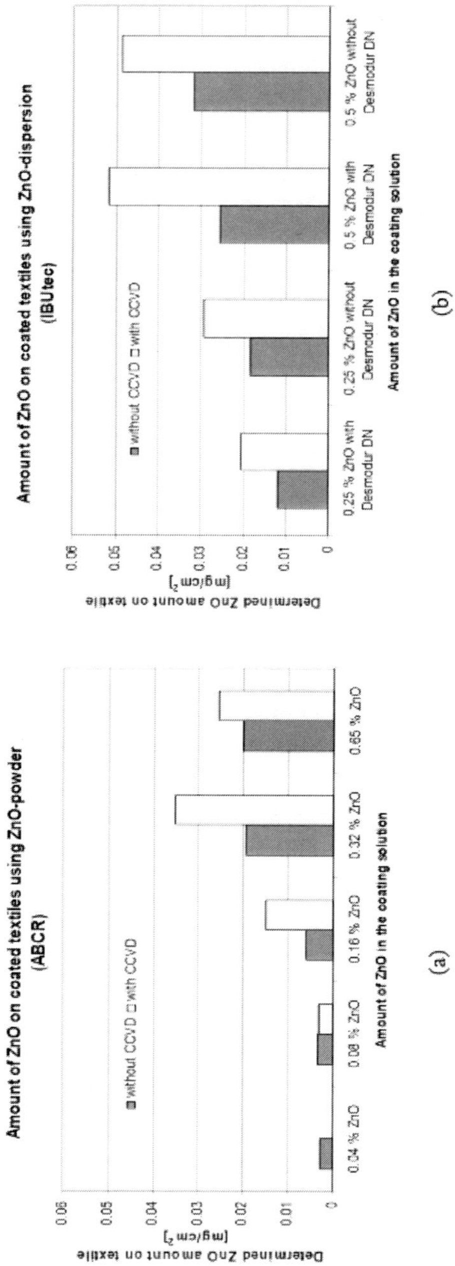

Figure 2. Measured amount of ZnO on industrial coated textiles using ZnO powder from ABCR and the binder Impranil DLU (a) as well as ZnO dispersion from IBUtec and Binder Baybond PU (b).

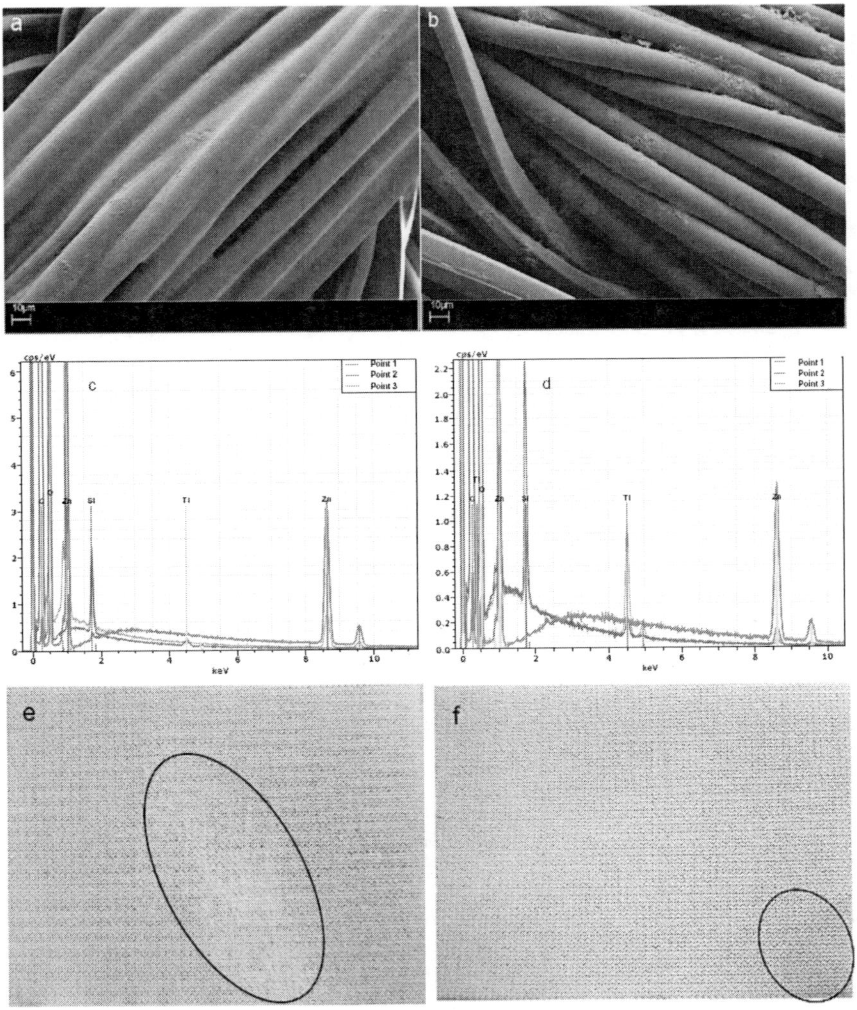

Figure 3. SEM images, EDX spectra (3 measurement points/sample) and photographic images of laboratory coated polyester samples with ZnO nanoparticles obtained from ABCR (a) (c) (e) and IBUtec (b) (d) (f).

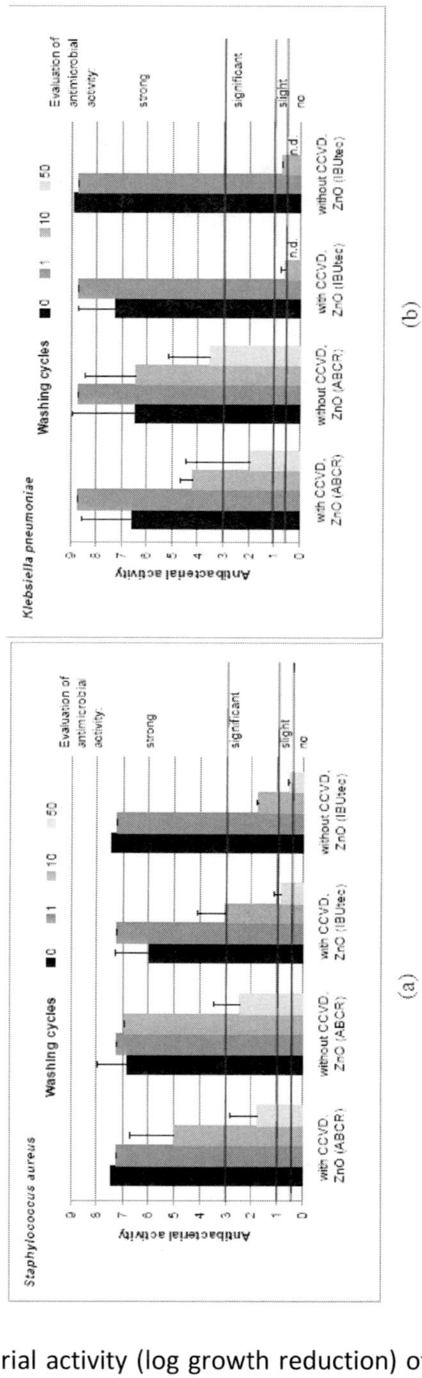

Figure 4. Antibacterial activity (log growth reduction) of laboratory coated and washed polyester textiles with ZnO and Impranil DLU, comparison of ABCR and IBUtec-particles (n. d. = not determined).

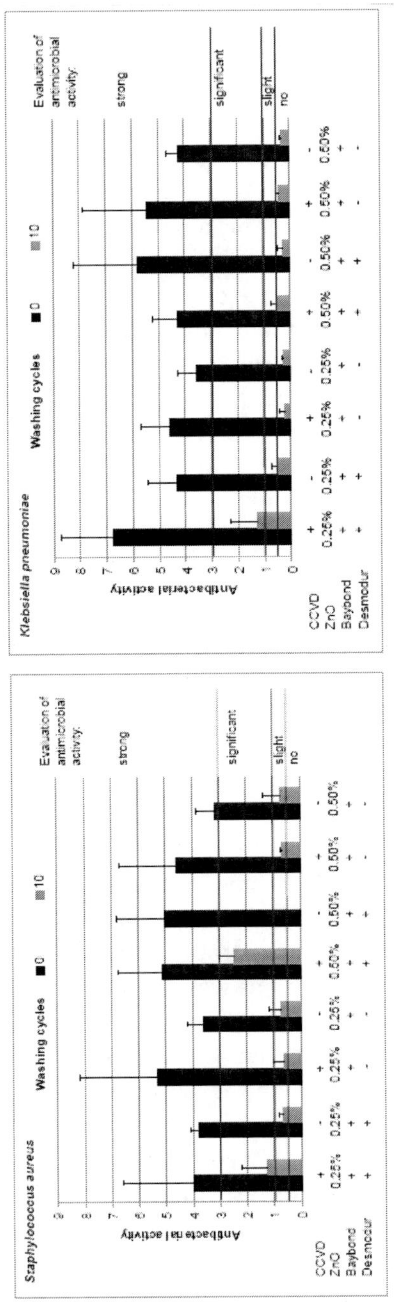

Figure 5. Antibacterial activity (log growth reduction) of industrial coated and washed polyester textiles with aqueous ZnO dispersion of IBUtec, Baybond PU and cross linking hardener Desmodur DN.

Results shown here are exemplarily for the concentrations of 0%, 0.25% and 0.5% ZnO (Figure 6). After an incubation period of 1 and 4 day, respectively, a large number of viable cells and only few dead ones could be detected on the control without coating. As can be seen on Figure 6(c), e that there are only a few viable cells after 1 day seeding on coated textile samples. At a concentration of 0.5% ZnO in the coating solution (amount of about 50 µg ZnO/cm^2 textile) there is a strong cytotoxicity after incubation of 1 day with 45% of dead cells and 4 days with 23% of dead cells. The rising number of viable cells causes the decreasing number of dead cells. By changing the concentration of ZnO to 0.25% (amount of about 20 µg ZnO/cm^{2}textile) we found cytotoxic effects only after 1 day of incubation with an amount of 22% of dead cells. After 4 days of seeding with 3T3 fibroblast cells we found no cytotoxic effects of the coating with 3% of dead cells (Figure 6(f)). Nevertheless the total number of viable cells is 65% in comparison with the uncoated textile sample. Investigations of Punnoose et al. revealed cytotoxicity of ZnO to prokaryotic and eukaryotic cells in higher concentrations [26] . It could be shown that the lethal concentration for zebra fish embryos is in a range of 50 - 100 mg/L ZnO nanoparticles and a lower concentration (1 - 25 mg/L) was accompanied with a retarded hatching of the embryos [27].

4. CONCLUSION

The antibacterial efficacy and washing stability of coated polyester samples depends on the composition of the coating solution. Particles of ZnO in the range of 50 - 300 nm demonstrate better results with regard to antibacterial efficacy after different washing cycles whereas nanoparticles with a size of 10 nm show improved optical appearance. Using the binder Impranil DLU for coating, slightly enhanced activity against Staphylococcus aureus and Klebsiella pneumoniae is obtained compared to the binder Baybond PU. Coating on an industrial scale with a foulard machine requires a finely dispersed ZnO solution for coating to avoid sedimentation of the nanoparticles. Because of the incompatibility of the zinc oxide dispersion with Impranil DLU the binder Baybond PU represents a favorable alternative for the developed coating system. Pretreatment of the textiles with CCVD-technique to generate a hydrophilic nano layer of hydroxyl terminated silicon network on the textile sur- face has no influence on the antibacterial efficacy and washing stability of the coating. But using CCVD the applied amount of ZnO can be increased on a foulard machine emphasizing the use of CCVD for industrial scale.

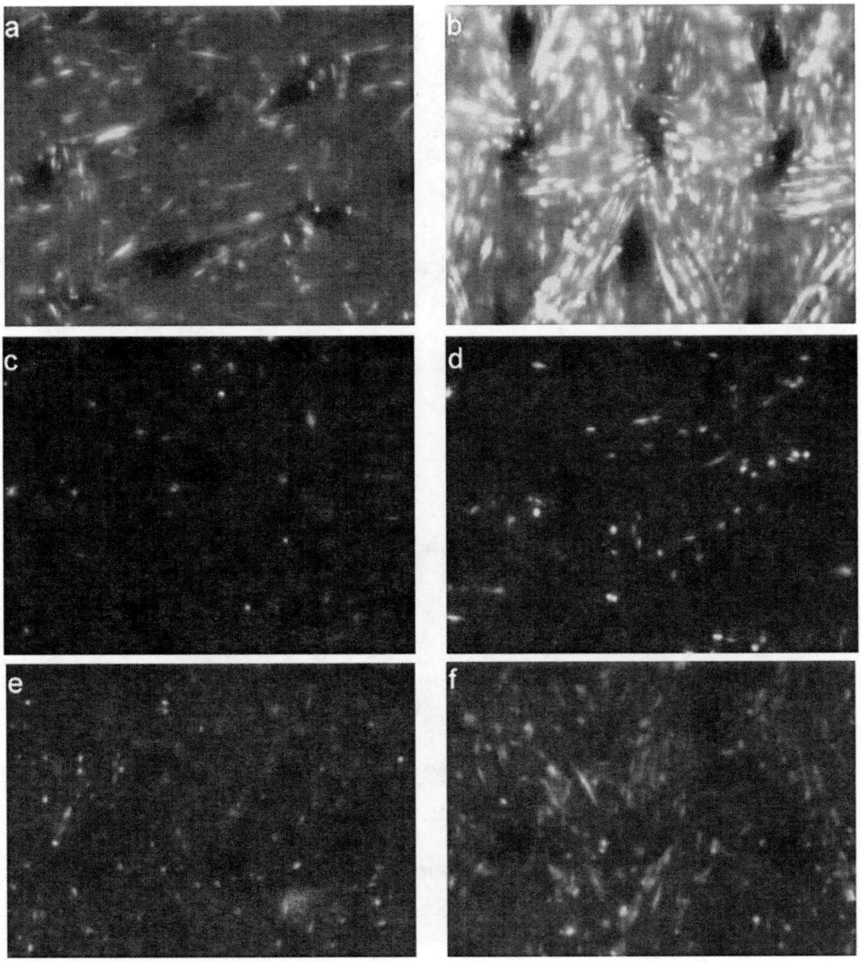

Figure 6. Cell viability testing of coated polyester samples using a FDA/GelRed® live/dead assay (a) (c) (e) cell incubating for 1 day; (b) (d) (f) cell incubating for 4 days; green fluorescence: living cells; red fluorescence: dead cell nuclei); (a) (b) polyester control without coating; (c) (d) coating with solution of 0.5% ZnO (IBUtec dispersion, amount of about 50 μg ZnO/cm^2 textile); (e) (f) coating with solution of 0.25% ZnO (IBUtec dispersion, amount of about 20 μg ZnO/cm^2 textile).

A ZnO concentration up to 20 μg ZnO/cm^2 is found to exhibit suitable antibacterial effect and requires cytocompatibility. Further studies in order to enhance washing stability of the antibacterial coated textiles have do be done.

ACKNOWLEDGEMENTS

We thank Dr. M. Schweder for SEM and EDX measurements and U. Gitter for the CCVD-treatment.

REFERENCES

1. Hsu, B.B. and Klibanov, A.M. (2011) Light-Activated Covalent Coating of Cotton with Bactericidal Hydrophobic Polycations. Biomacromolecules, 12, 6-9.

2. Liang, J., Chen, Y., Barnes, K., Wu, R., Worley, S.D. and Huang T.S. (2006) N-Halamine/Quat Siloxane Copolymers for Use in Biocidal Coatings. Biomaterials, 27, 2495-2501.

3. Gao, Y. and Cranston, R. (2008) Recent Advances in Antimicrobial Treatments of Textiles. Textile Research Journal, 78, 60-72.

4. Mahltig, B., Fiedler, D. and B&oumlttcher, H. (2004) Antimicrobial Sol-Gel Coatings. Journal of Sol-Gel Science and Tech- nology, 32, 219-222.

5. Radetic, M., Ilic, V., Vodnik, V., Dimitrijevic, S., Jovancic, P., Saponjic, Z. and Nedeljkovic, J.M. (2008) Antibacterial Effect of Silver Nanoparticles Deposited on Corona-Treated Polyester and Polyamide Fabrics. Polymers for Advanced Technologies, 19, 1816-1821.

6. Ilic, V., Saponjic, Z., Vodnik, V., Lazovicic, S., Dimitrijevic, S., Jovancic, P., Nedeljkovic, J.M. and Radetic, M. (2010) Bactericidal Efficiency of Silver Nanoparticles Deposited onto Radio Frequency Plasma Pretreated Polyester Fabrics. Industrial & Engineering Chemistry Research, 49, 7287-7293.

7. Mejía, M.I., Restrepo, G., Marín, J.M., Sanjines, R., Pulgarín, C., Mielczarski, E., Mielczarski, J. and Kiwi, J. (2010) Magnetron-Sputtered

Ag Surfaces. New Evidence for the Nature of the Ag Ions Intervening in Bacterial Inactivation. ACS Applied Materials & Interfaces, 2, 230-235.

8. Torres, A., Ruales, C., Pulgarin, C., Aimable, A., Bowen, P., Sarria, V. and Kiwi, J. (2010) Innovative High-Surface- Area CuO Pretreated Cotton Effective in Bacterial Inactivation under Visible Light. ACS Applied Materials & Interfaces, 2, 2547-2552.

9. Chen, Z., Luo, J. and Sun, Y. (2007) Biocidal Effiacy, Biofilm-Controlling Function, and Controlled Release Effect of Chloromelamine-Based Bioresponsive Fibrous Materials. Biomaterials, 28, 1597-1609.

10. Ringot, C., Sol, V., Barrière, M., Saad, N., Bressollier, P., Granet, R., Couleaud, P., Frochot, C. and Krausz, P. (2011) Triazinyl Porphyrin-Based Photoactive Cotton Fabrics: Preparation, Characterization, and Antibacterial Activity. Biomacromolecules, 12, 1716-1723.

11. El-tahlawy, K.F., El-bendary, M.A., Elhendawy, A.G. and Hudson, S.M. (2005) The Antimicrobial Activity of Cotton Fabrics Treated with Different Crosslinking Agents and Chitosan. Carbohydrate Polymers, 60, 421-430.

12. Alonso, D., Gimeno, M., Olayo, R., Vázquez-Torres, H., Sepúlveda-Sánchez, J.D. and Shirai, K. (2009) Cross-Linking Chitosan into UV-Irradiated Cellulose Fibers for the Preparation of Antimicrobial-Finished Textiles. Carbohydrate Polymers, 77, 536-543.

13. Brausch, J.M. and Rand, G.M. (2011) A Review of Personal Care Products in the Aquatic Environment: Environmental Concentrations and Toxicity. Chemosphere, 82, 1518-1532.

14. Holdsworth, S.R. and Law, C.J. (2013) The Major Facilitator Superfamily Transporter MdtM Contributes to the Intrinsic Resistance of Escherichia coli to Quaternary Ammonium Compounds. Journal of Antimicrobial Chemotherapy, 68, 831-839.

15. Braga, T.M., Marujo, P.E., Pomba, C. and Lopes, M.F.S. (2011) Involvement, and Dissemination, of the Enterococcal Small Multidrug Resistance Transporter QacZ in Resistance to Quaternary Ammonium Compounds. Journal of Antimicrobial Chemotherapy, 66, 283-286.

16. Ciusa, M.L., Furi, L., Knight, D., Decorosi, F., Fondi, M., Raggi, C., Coelho, J.R., Aragones, L., Moce, L., Visa, P., Freitas, A.T., Baldassarri, L., Fani,

R., Viti, C., Orefici, G., Martinez, J.L., Morrissey, I. and Oggioni, M.R. (2012) A Novel Resistance Mechanism to Triclosan That Suggests Horizontal Gene Transfer and Demonstrates a Potential Selective Pressure for Reduced Biocide Susceptibility in Clinical Strains of Staphylococcus aureus. International Journal of Antimicrobial Agents, 40, 210-220.

17. Vigneshwaran, N., Kumar, S., Kathe, A.A., Varadarajan, P.V. and Prasad, V. (2006) Functional Finishing of Cotton Fabrics Using Zinc Oxide-Soluble Starch Nanocomposites. Nanotechnology, 17, 5087-5095.

18. Rajendran, R., Balakumar, C., Ahammed, H.A.M., Jayakumar, S., Vaideki, K. and Rajesh, E. (2010) Use of Zinc Oxide Nano Particles for Production of Antimicrobial Textiles. International Journal of Engineering Science and Technology, 2, 202-208.

19. Kathirvelu, S., D'Souza, L. and Dhurai, B. (2008) A Comparative Study of Multifunctional Finishing of Cotton and P/C Blended Fabrics Treated with Titanium Dioxide/Zinc Oxide Nanoparticles. Indian Journal of Science and Tech- nology, 1, 1-12.

20. Liu, Y., He, L., Mustapha, A., Li, H., Hu, Z.Q. and Lin, M. (2009) Antibacterial Activities of ZnO Nanoparticles against Escherichia coli O157:H7. Journal of Applied Microbiology, 107, 1193-1201.

21. Xie, Y., He, Y., Irwin, P.L., Jin, T. and Shi, X. (2011) Antibacterial Activity and Mechanism of Action of Zinc Oxide Nanoparticles against Campylobacter jejuni. Applied and Environmental Microbiology, 77, 2325-2331.

22. Nair, S., Sasidharan, A., Divya Rani, V.V., Menon, D., Nair, S., Manzoor, K. and Raina, S. (2009) Role of Size Scale of ZnO Nanoparticles and Microparticles on Toxicity toward Bacteria and Osteoblast Cancer Cells. Journal of Materials Science: Materials in Medicine, 20, 235-241.

23. Brayner, R., Ferrari-Iliou, R., Brivois, N., Djediat, S., Benedetti, M.F. and Fievet, F. (2006) Toxicological Impact Studies Based on Escherichia coli Bacteria in Ultrafine ZnO Nanoparticles Colloidal Medium. Nano Letters, 6, 866- 870.

24. Wiegand, C., Heinze, T. and Hipler, U.C. (2009) Comparative in Vitro Study on Cytotoxicity, Antimicrobial Activity, and Binding Capacity for

Pathophysiological Factors in Chronic Wounds of Alginate and Silver-Containing Alginate. Wound Repair and Regeneration, 17, 511-521.

25. Zhang, L., Jiang, Y., Ding, Y., Povey, M. and York, D. (2007) Investigation into the Antibacterial Behaviour of Suspensions of ZnO Nanoparticles (ZnO Nanofluids). Journal of Nanoparticle Research, 9, 479-489.

26. Reddy, K.M., Feris, K., Bell, J., Wingett, D.G., Hanley, C. and Punnoose, A. (2007) Selective Toxicity of Zinc Oxide Nanoparticles to Prokaryotic and Eukaryotic Systems. Applied Physics Letters, 90, Article ID: 213902.

27. Bai, W., Zhang, Z., Tian, W., He, X., Ma, Y., Zhao, Y. and Chai, Z. (2010) Toxicity of Zinc Oxide Nanoparticles to Zebrafish Embryo: A Physicochemical Study of Toxicity Mechanism. Journal of Nanoparticle Research, 12, 1645- 1654.

Chapter 11

Synthesis of Some Novel 2-Amino-5-arylazothiazole Disperse Dyes for Dyeing Polyester Fabrics and Their Antimicrobial Activity

Hatem E. Gaffer [1,†], **Moustafa M. G. Fouda** [1,2,†]
and Mohamed E. Khalifa [3,4,*]

[1]Textile Research Division, National Research Center, Dokki, Giza, P. O. Box.12611, Egypt

[2]Chemistry Department, College of Science, King Saud University, P. O. Box 2455, Riyadh 11451, Saudi Arabia

[3]Chemistry Department, Faculty of Science, Taif University, P. O. Box 888, Taif 21974, Saudi Arabia

[4]Department of Chemical Engineering, Higher Institute for Engineering and Technology, New Damietta, P. O. Box 34518, Egypt

ABSTRACT

The present work describes the synthesis of a series of four novel biologically active 2-amino-5-arylazothiazole disperse dyes containing the sulfa drug nucleus. The structures of the synthesized thiazole derivatives are confirmed using UV-spectrophotometry, infrared and nuclear magnetic resonance techniques and elemental analysis. The synthesized dyes are applied to polyester fabrics as disperse dyes and their fastness properties to washing, perspiration, rubbing, sublimation, and light are evaluated. The synthesized compounds exhibit promising biological efficiency against selected Gram-positive and Gram-negative pathogenic bacteria as well as fungi.

KEYWORDS

azo dyes; 2-aminothiazole; sulfaguanidine; polyester; dyeing; antimicrobial activity

1. INTRODUCTION

Azo dyes are the most essential group of disperse dyes [1,2] and universally constitute more than 50% of disperse dyes due to their strong tinctorial strength, ease of synthesis and low cost of manufacture. Azo dyes have vivid colors in the yellow to blue-green region adjustable by varying between azo components with electron withdrawing groups and coupling components with electron donating groups. 2-Aminothiazoles have considerable interest from a therapeutic point of view because of their utility as intermediates in the synthesis of antibiotics such as the well-known sulfa drugs [3]. They have been reported in the literature as antimicrobial [4], antifungal [5], anti-inflammatory activity [6], anesthetic [7], anti-hypertesive [8] and antiviral drugs [9]. 2-Aminothiazole derivatives have been also used as intermediates in the synthesis of various types of dyes [10,11,12,13]. In continuation of our interest in the synthesis of arylazothiazole derivatives [14,15,16], this paper reports the synthesis of some 2-amino-5-arylazothiazoles and their application as disperse dyes on polyester fabrics. In addition, the antibacterial activities of the synthesized dyes against various pathogenic bacteria and fungi were also investigated.

2. RESULTS AND DISCUSSION

2.1. Synthesis

Treatment of 2-aminothiazole derivatives **1** with the diazonium chloride derived from sulfaguanidine **2** afforded the corresponding 2-amino-5-

arylazothiazole derivatives **4** and **5** as shown in Scheme 1. The structures of **4** and **5** were established on the basis of their elemental analyses and spectral data. For example, the IR spectrum of **4** revealed absorption bands at 3346, 3232, 3173 cm^{-1} due to NH_2 and NH functions, while the absorption band at 1610 cm^{-1} was attributed to the C=N function. The ^{1}H-NMR spectrum displayed a triplet signal at δ 7.05 ppm due to the NH proton, two doublet signals at δ 7.74 and 8.04 ppm corresponding to the aromatic protons, a singlet at δ 8.16 ppm characteristic for the C-4 thiazole-H, two signals at δ 8.51 and 8.67 due to two NH_2 groups and a singlet signal at δ 11.88 ppm due to NH proton.

Moreover, coupling of 2-aminothiazole derivatives **1** with the diazonium chloride derived from sulfapyrimidine **3** furnished the corresponding 2-amino-5-arylazothiazole derivatives **6** and **7** as shown in Scheme 1. The structures of these substituted thiazole dyes **6** and **7** were elucidated based on their elemental analyses and spectral data. The IR of the dye **6** showed absorption peaks at 3364, 3286, 3142 cm^{-1} due to the presence of NH_2 and NH functions, in addition to the characteristic C=N group absorption peak at 1612 cm^{-1}. The ^{1}H-NMR spectrum of dye **6** showed a triplet signal at 6.88 ppm due to C-5 pyrimidine proton, a singlet at 7.13 ppm for the NH group, two doublets at 7.66 ppm and 8.04 ppm for four aromatic protons and a singlet at 8.22 ppm for the C-4 thiazole proton. The presence of a doublet at 8.40 was attributed to the C-4 and C-6 pyrimidine-H protons, while the NH_2 group appeared shifted to 8.70 ppm.

2.2. Dyeing and Fastness Properties

The functionalized 4-arylazo-3-methyl-2-substituted-thiophene disperse dyes **4–7** were applied to polyester fabrics by 2% shade on the weight of fabric (o.w.f.), using the high-temperature pressure technique (130 °C), and the visual color shades varied from red to reddish violet. Dyeing of polyester fabrics was assessed in terms of their fastness properties (e.g., fastness to washing, perspiration, rubbing, sublimation, and light) via a standard method [17]. All results, provided inTable 1, reveal that these dyes have good fastness properties to washing, perspiration, rubbing and sublimation.

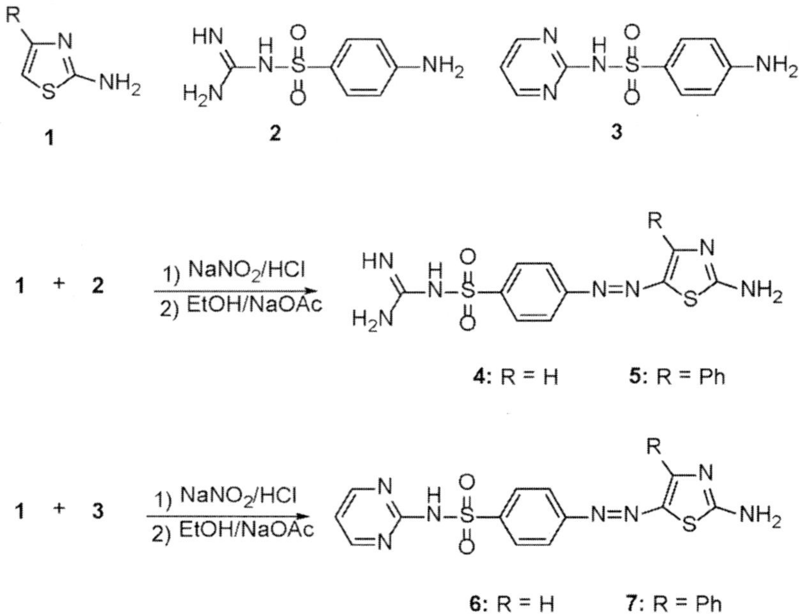

Scheme 1. Synthesis of 2-amino-4-arylazo-thiazole disperse dyes **4–7**.

The light fastness of the synthesized dyes on polyester showed fading which is significantly affected by the presence of the azo group, which decomposes by oxidation, reduction and/or photolysis [18]. The rate of this process depends on the chemical structure, treatment conditions, as well as the nature of the substrate. The use of polyester fabric (*i.e.*, non-proteinic fabric type), explained that the fading process likely occurred by oxidation [19]. The ease of oxidation of azo linkages should be a function of their electron density, thus the presence of electron-donating substituents should increase the fading rate. This proposal was in agreement with the observed results (Table 1) which demonstrate that the presence of the attached phenyl group decreased of light fastness to 4–5 (dyes **5** and **7**).

Table 1. Fastness properties of the synthetic dyes on polyester fabrics.

Dye	Washing [1]	Perspiration [1]		Rubbing [1]		Sublimation [1]		Light [2] (60 h)
		Acid	Alkali	Dry	Wet	Staining at 180 °C	Staining at 210 °C	
4	4–5	4–5	4–5	3–4	4–5	4	3	5
5	4–5	3–4	4–5	3	3	5	4	4–5
6	4–5	4–5	4–5	3–4	4–5	5	4	5
7	4–5	4	4	4	4	3	3	4–5

[1] These properties were evaluated using the grey scale (1—poor, 2—fair, 3—moderate, 4—good and 5—excellent).
[2] This property was evaluated using the blue scale (1, 2—very poor, 3—poor, 4—fair, 5—good, 6, 7—very good, 8—excellent) [17].

2.3. Color Assessment

Colors of polyester fabrics is expressed in terms of CIELAB values (Table 2) and the following CIELAB coordinates are assessed: lightness (L*), chroma (C*), hue angle from 0° to 360° (H), (a*) value exemplifies the degree of redness (positive) and greenness (negative) and (b*) signifies the degree of yellowness (positive) and blueness (negative). A reflectance spectrophotometer (Gretag Macbeth CE 7000a) was utilized for the colorimetric assessments of the dyed samples. K/S values provided by the reflectance spectrometer were calculated at λ_{max} and were directly associated with the dye concentration on the dye substrate.

Generally, the color hues of the thiazole dyes **5** and **7** on polyester fabric shifted to the greenish directions, this was expressed by the negative values of a* (red-green axis). The positive values of b* (yellow-blue axis) designated that the color hues of the thiazole dyes **4–7** on polyester fabric are shifted to the yellowish directions as shown in Table 2.

Application of the synthesized compounds **4–7** as disperse dyes, showed the good affinity of such dyes to polyester fabrics, as indicated from the satisfactory color yields and the acceptable K/S values. Figure 1, illustrated the relationship between the dyeing bath concentrations and the strengths of the synthesized disperse dyes **4–7**, under high temperature dyeing conditions, where no direct connection between the relative molecular weight of these dyes to the build-up was clear. These results were in line with the previously reported by Müller [20] on the relation between substituents and dye structure.

2.4. Antimicrobial Activity

Table 3 shows that all samples exhibited an acceptable antimicrobial activity against the tested pathogenic bacteria at different concentrations 0.01, 0.5 and 1 μg/mL, where all the results are mean of three replicates and significant difference at $p < 0.05$ by student's test. At high concentrations, regarding *C. albicans*; however, the extract exhibited an acceptable antifungal activity. The inhibition zone of all synthetic products was at range 0.6 to 1.7 mm. The remarkable activity of the synthesized compounds **4–7** against Gram-negative bacteria, which were known to be more resistant against antibiotics than Gram-positive bacteria because of their impenetrable cell wall, should be undergo further studies in order to understand well the mechanism of action for each active compound.

Table 2. Optical measurements of the synthetic dyes on polyester fabrics.

Dye	%Exhaustion	Absorption λ_{max}/nm	K/S	L*	a*	b*	C*	H
4	65	446	9.26	81.09	3.35	20.42	40.58	80.71
5	71	444	12.67	73.23	−1.48	23.13	43.88	76.75
6	69	463	11.33	74.22	8.17	37.20	48.71	76.21
7	66	458	9.37	89.35	−1.78	3.14	43.51	85.84

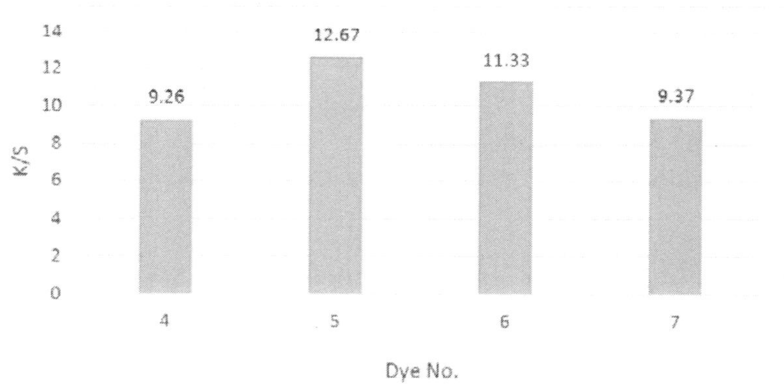

Figure 1. Comparison of K/S values for the synthesized disperse dyes **4**–**7** series.

2.5. Scanning Electron Microscopy

Scanning electron microscopy was used to observe the morphological alteration of the cells after treatment with the synthesized compounds (e.g., the most active compound **7**). As shown in Figure 2 and Figure 3, the bacterial (*S. aureus*) and fungal (*Candida albicans*) cells treated with the synthesized compound appeared to be shrink and the cell wall degradation was observed.

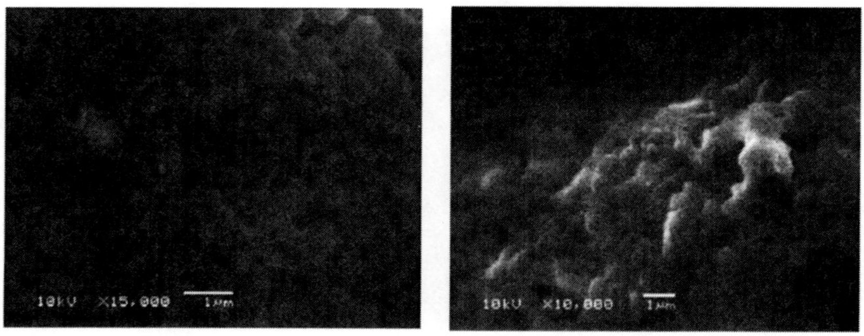

Figure 2. *S. aureus* in ethanol as control (**left**); *S. aureus* treated with synthesized compound **7** (**right**).

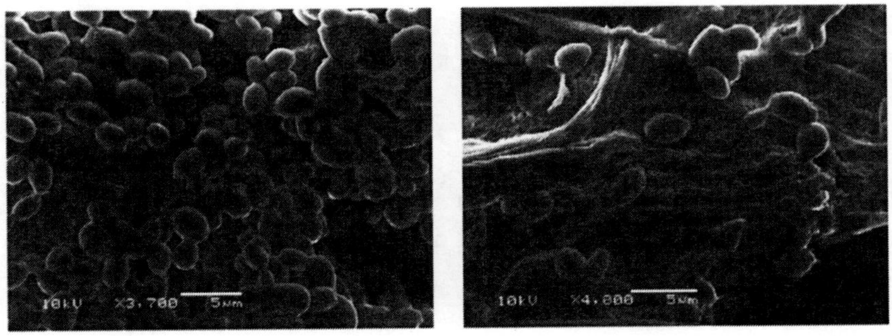

Figure 3. *Candida albicans* in ethanol as control (**left**); *Candida albicans* treated with synthesized compound **7** (**right**).

2.6. Minimum Inhibitory Concentration (MIC)

A strong inhibition of the synthesized compounds **4**–**7** against the tested pathogenic bacteria was noticed (5 to 150 μg/mL) as described in Table 4, to be vulnerable to all of them and their minimum inhibition concentrations (MIC) values were ranged from 5–10 μg/mL, with noticeably inhibitory action. In this study, synthesized compounds **4** and **7** showed the highest antibacterial activity against the tested bacteria with the lowest MIC value of 5 μg/mL. According to the obtained results, it is obviously clear that, all synthesized compounds **4**–**7** exhibited broad-spectrum activity against both Gram-positive bacteria, Gram-negative bacteria as well as fungi.

3. EXPERIMENTAL SECTION

3.1. Materials and Methods

3.1.1. Materials, Chemicals and Reagents

All the chemicals and solvents used in this study were obtained from Acros Organics, Thermo Fisher Scientific Company (New Jersey, NJ, USA), and of pure grades. Polyester fabrics were kindly supplied by Misr for Spinning and Weaving Company (Mahalla El-Kobra, Egypt). The dispersing agent Setamol WS was supplied by BASF (Nienburg, Germany). The bacterial strains from the American Type Culture Collection (ATCC) were purchased from

MicroBioLogics Inc., (St. Cloud, MN, USA). The tested microorganism strains in ethanol were used as controls for biological experiments.

Table 3. Antimicrobial activity expressed as inhibition diameter zones (mm) of the synthesized compounds **4–7** at different concentrations (μg/mL).

Compd. No.	Gram-Positive Bacteria						Gram-Negative Bacteria									Fungi	
	S. aureus			*S. pyogenes*			*S. typhi*			*P. aeruginosa*			*K. pneumoniae*			*C. albicans*	
	0.01	0.5	1.0	0.01	0.5	1	0.01	0.5	1	0.01	0.5	1	0.01	0.5	1.0	0.005	0.05
4	0.6	0.7	1	0.4	0.6	0.65	0.6	0.9	1.2	0.3	0.6	0.8	0.2	0.6	0.7	0.7	1.1
5	0.6	0.8	1.2	0.1	0.2	0.6	0.6	0.8	1	0.2	0.4	0.6	0.2	0.6	0.7	0.9	1.3
6	0.6	1	1.7	0.3	0.6	1.6	0.6	1	1.2	0.7	1	1.4	0.3	0.5	1.2	0.7	1.4
7	0.6	1.2	1.3	0.6	1.1	1.9	0.7	1.2	2	0.1	0.3	0.7	0.4	1	2	0.8	1.3
Ciprofloxacin	0.9	1.2	1.4	0.8	1.0	1.3	0.8	0.9	1.5	0.8	1.0	1.4	0.5	0.8	1.2	-	-
Ketoconazole	-	-	-	-	-	-	-	-	-	-	-	-	-	-	-	0.8	1.4

Table 4. Minimum Inhibitory Concentration (µg/mL) of the synthesized compounds **4–7** against Gram +ve (e.g., *S. aureus*and Gram -ve (e.g., *S. typhi*) bacteria at different concentrations (µg/mL).

Compd.#	Control Conc. (%)	*S. aureus*				Control Conc. (%)	*S. typhi*			
		5	10	60	150		5	10	60	150
4	49.32×10^3 (0 %)	37.65×10^3 (23.7 %)	31.44×10^3 (36.3 %)	24.80×10^3 (49.7 %)	19.39×10^3 (60.7 %)	36.09×10^3 (0 %)	34.22×10^3 (5.2 %)	25.36×10^3 (29.7 %)	22.71×10^3 (37.1 %)	16.64×10^3 (53.9 %)
5	49.32×10^3 (0 %)	43.47×10^3 (11.9 %)	26.16×10^3 (47.0 %)	25.22×10^3 (48.9 %)	17.32×10^3 (64.9 %)	36.09×10^3 (0 %)	36.56×10^3 (1.3 %)	31.89×10^3 (11.6 %)	22.24×10^3 (38.4 %)	15.56×10^3 (56.9 %)
6	49.32×10^3 (0 %)	48.94×10^3 (0.8 %)	48.19×10^3 (1.0 %)	46.31×10^3 (6.1 %)	28.24×10^3 (42.7 %)	36.09×10^3 (0 %)	32.67×10^3 (9.5 %)	20.22×10^3 (44.0 %)	19.44×10^3 (46.1 %)	16.18×10^3 (55.2 %)
7	49.32×10^3 (0 %)	48.21×10^3 (21.2 %)	37.65×10^3 (23.7 %)	20.52×10^3 (58.4 %)	19.20×10^3 (61.1 %)	36.09×10^3 (0 %)	31.89×10^3 (11.6 %)	24.00×10^3 (33.5 %)	21.78×10^3 (39.7 %)	16.33×10^3 (54.8 %)

3.1.2. Instrumentation

All melting points were measured by the capillary method on an electrothermal Gallenkamp melting point apparatus (Gallenkamp Co., London, UK) and were uncorrected. Elemental analyses were carried out at the Microanalytical Unit, National Research Centre, Giza, Egypt; the results were in satisfactory agreement with the calculated values. The IR spectra (KBr) were recorded in KBr disks by a Mattson 5000 FTIR spectrometer (Mattson Inst. Co., Madison, WI, USA) and not all frequencies are reported. The NMR spectra (^{1}H and ^{13}C) were acquired using a WP 300 spectrometer (Bruker Co., Billerica, MA, USA) at 300 MHz using TMS as an internal standard and DMSO-d_6 as solvent. The colorimetric measurements for the dyed polyester fabrics were carried out using a reflectance spectrophotometer (GretagMacbeth CE 7000a, GretagMacbeth Co., Windsor, UK). Fastness to washing was carried out using the automatic Rotadyer launder (sponsored by the British Standard Institute—Society of Dyers and Colourists), fastness to perspiration was assessed according to the test sponsored by the (BSS), fastness to rubbing was carried out according to the standard method of testing (BSS) using a Crockmeter FD-17 apparatus (Electric Hungarian Co., Budapest, Hungary), fastness to sublimation was carried out using scorch tester M247 A (Atlas Electric Devices Co., Chicago, IL, USA) and fastness to light was carried out using the "Weather-o-meter" (Atlas Electric Devices Co.) according to the AATCC standard test method.

3.2. Synthesis

3.2.1 General Procedure for the Synthesis of 2-Amino-4-arylazo-thiazole Disperse Dyes 4–7

A solution of aryl diazonium chloride was prepared by adding cold sodium nitrite solution (0.7 g in 10 mL H_2O) to a cold suspension of aryl amine **2** and/or **3** (0.01 mol) in 3 mL concentrated HCl with stirring. The freshly prepared solution was added with continuous stirring to a cold solution (0–5 °C) of 2-aminothiazole **1** (1 g, 0.01 mol) in 30 mL ethanol and 4.0 g sodium acetate. The reaction mixture was stirred at 0–5 °C for 2 h, diluted with water, and the products collected by filtration. The obtained 2-amino-5-arylazothiazoles **4–7** were dried and recrystallized from ethanol.

2-Amino-5-{[4-(carbamimidoyl-sulfamoyl)-phenyl]azo}thiazole (**4**). Red solid, yield 74%, m.p. 220–221 °C. IR (KBr, cm^{-1}): 3432, 3312, 3177 (NH_2 and NH), 1640 (C=N). ^{1}H-NMR (DMSO-d_6, δ ppm): 7.05 (t, 1H, J = 7.15 Hz, NH), 7.59 (d, 2H, J = 7.80 Hz, Ar-H), 7.97 (d, 2H, J = 7.80 Hz, Ar-H), 8.36 (s, 1H, C-4 thiazole-H), 8.65 (d, 2H, J = 8.50 Hz, NH_2), 9.40 (s, 2H, NH_2), 11.35 (s, 1H, NH). ^{13}C-NMR (DMSO-d_6, δ ppm): 121.99 (2C), 129.51 (2C), 139.79, 144.88, 152.80, 155.20, 158.91, 173.40. Anal. Calcd. for $C_{10}H_{11}N_7O_2S_2$ (325.37): C, 36.91; H, 3.41; N, 30.13,Found: C, 36.83; H, 3.45; N, 30.20.

2-Amino-5-{[4-(carbamimidoyl-sulfamoyl)-phenyl] azo}-4-phenylthiazole (**5**). Red solid, yield 77%, m.p. 241–241 °C. IR (KBr, cm^{-1}): 3433, 3393, 3321 (NH_2 and NH), 1639 (C=N). ^{1}H-NMR (DMSO-d_6, δ ppm): 6.94 (s, 1H, NH), 7.15–7.50 (m, 7H, Ar-H), 7.78 (d, 2H, J = 7.80 Hz, Ar-H), 8.48 (s, 2H, NH_2), 9.18 (s, 2H, NH_2), 11.37 (s, 1H, NH). ^{13}C-NMR (DMSO-d_6, δ ppm): 121.46 (2C), 126.55 (2C), 128.21, 129.38 (2C), 130.86 (2C), 137.32, 140.64, 144.32, 151.08, 154.56, 160.27, 174.38. Anal. Calcd. For $C_{16}H_{15}N_7O_2S_2$ (401.47): C, 47.87; H, 3.77; N, 24.42, Found: C, 47.73; H, 3.84; N, 24.50.

2-Amino-5-{[4-(pyrimidin-2-yl-sulfamoyl)-phenyl] azo}thiazole (**6**). Red solid, yield 80%, m.p. 234–235 °C. IR (KBr, cm^{-1}): 3351, 3254, 3146 (NH_2 and NH), 1625 (C=N). ^{1}H-NMR (DMSO-d_6, δ ppm): 6.80 (s, 1H, NH), 7.17 (d, 1H, J = 7.50 Hz, C-5 pyrimidine-H), 7.32 (d, 2H, J = 7.80 Hz, Ar-H), 7.63 (d, 2H,J = 7.80 Hz, Ar-H), 7.96 (d, 2H, J = 8.80 Hz, C-4 and C-6 pyrimidine-H), 8.25 (s, 1H, C-4 thiazole-H), , 9.79 (s, 2H, NH_2). ^{13}C-NMR (DMSO-d_6, δ ppm): 112.82, 122.48 (2C), 127.75 (2C), 131.84, 138.79, 140.11, 142.68, 156.21 (2C), 167.86, 173.36. Anal. Calcd. For $C_{13}H_{11}N_7O_2S_2$ (361.40): C, 43.20; H, 3.07; N, 27.13, Found: C, 43.38; H, 3.15; N, 27.24.

2-Amino-4-phenyl-5-{[4-(pyrimidin-2-yl-sulfamoyl)-phenyl]azo}thiazole (**7**). Red solid, yield 72%, m.p. 181–182 °C. IR (KBr, cm^{-1}): 3463, 3372, 3348, 3234 (NH_2 and NH), 1625 (C=N). ^{1}H-NMR (DMSO-d_6, δ ppm): 6.84 (d, 1H, J = 7.50 Hz, C-5 pyrimidine-H), 7.11 (s, 1H, NH), 7.27–7.61 (m, 11H, Ar-H), 7.83 (d, 2H, J = 8.80 Hz, C-4 and C-6 pyrimidine-H), 8.62 (d, 2H, J = 8.50 Hz, NH_2). ^{13}C-NMR (DMSO-d_6, δ ppm): 113.64, 123.71 (2C), 127.06 (2C), 128.46, 129.18 (2C), 130.69 (2C), 135.74, 136.62 138.08, 141.26, 144.59, 157.27 (2C), 167.38, 172.91.Anal. Calcd. For $C_{19}H_{15}N_7O_2S_2$ (437.50): C, 52.16; H, 3.46; N, 22.41, Found: C, 52.02; H, 3.57; N, 22.57.

3.3. Disperse Dyeing Application

3.3.1. Dyebath Preparation Dyeing Procedure

The synthesized disperse dyes **4–7** under investigation were applied to polyester fabrics (0.04 g dye/2 g fabric; o.w.f. 2%) by a convenient method for dyeing polyester fabrics in the laboratory at 130 °C and high-pressure (24–30 psi). A dispersion of dye was formed by dissolving the proper quantity of dye in 1 mL of acetone as solvent. 1% Setamol WS was added dropwise with continuous stirring to the dye bath as anionic dispersing agent. The liquor ratio was 20:1, and the pH was adjusted to 5.5 using aqueous acetic acid. Then, the wetted-out polyester fabrics were added to the dyebath. Dyeing process was accomplished by rising the dyebath temperature to 130 °C, at time intervals of 3 °C/min, the dyeing process was left for 60 min. After cooling to 50 °C, the dyed fabrics were rinsed with cold water and then reduction cleared using a mixture of 1 g/L sodium hydroxide and 1 g/L sodium hydrosulfite for 10 min. at 80 °C. The dyed fabrics were finally rinsed again with hot water followed by cold water and left for air drying.

3.3.2. Dyeing Characteristics on Polyester Fabrics

3.3.2.1 Color Fastness Tests

The color fastness properties of the dyed fabrics to washing, perspiration, rubbing, sublimation and light were evaluated using standard methods [17]. The staining of adjacent cotton and nylon fabrics was assessed using the grey scale: 1—poor, 2—fair, 3—moderate, 4—good and 5—excellent, while the light fastness properties were scaled from 1–8 on the blue scale.

3.3.2.2 Color Properties of the Dyes on Polyester

A GretagMacbeth CE 7000a reflectance spectrophotometer (D65 illumination, 10° observer) was used for the colorimetric measurements on the dyed samples. K/S values given by the reflectance spectrometer were calculated at λ_{max} and were directly correlated with the dye concentration on the dye substrate according to the Kubelka–Munk equation: K/S = (1 –

$R)^2/2R$, where K = absorbance coefficient, S = scattering coefficient, R = reflectance ratio.

3.4. Biological Assessment

3.4.1. Test-Pathogens

The antibacterial activity tests were performed against five bacteria strains: Gram-positive (*Staphylococcus aureus* ATCC 25923 and *Streptococcus pyogenes* ATCC 19615) and Gram-negative (*Salmonella typhi* ATCC 6539, *Pseudomonas aeruginosa* ATCC 27853 and *Klebsiella pneumoniae* ATCC 700603), as well as *Candida albicans* ATCC 90028 for the antifungal activity test. The microorganisms under test were grown onto Sabouraud agar. The microbial strains were conserved in Brain Heart Infusion medium (BHI) at −20 °C. Three hundred μL of each stock-culture was added to 3 mL of BHI broth. Cultures were kept for 24 h at 37 °C ± 1 °C and the purity of cultures was checked after 8 h of incubation. After 24 h of incubation, bacterial suspension was diluted with sterile sodium chloride solution, and bacterial suspension was diluted with BHI broth to a density of approximately 10^9 colony forming unit/mL (CFU/mL) [21].

3.4.2. Antimicrobial and Antifungal Activity

The antimicrobial activities of the dyes **4–7** were assessed *in vitro* using the agar diffusion method [22]. In this method, the medium was incubated with freshly prepared cells for each tested bacteria or fungi. After solidification of the agar, a number of clean disks were immersed into the solvents (negative controls) or extract solutions and located on the plates. After incubation at 37 °C for 24 h, the antimicrobial activity was assessed in term of diameter of the inhibition zone. Each test was repeated three times and the average results (mm of zone of inhibition) is tabulated.

3.4.3. Minimum Inhibitory Concentration (MIC)

The selected bacteria for MIC were *Staphylococcus aureus* ATCC 25923 and *Salmonella typhi* ATCC 6539. MIC of four different samples was calculated by the two-fold serial dilution method [22,23]. The levels dose of 5, 10, 60, 150 μL/mL were utilized for MIC evaluation. 5 μL of standardized suspension of tested bacteria (10^8 CFU/mL) well was used for 1 mL of the broth. The test tubes were incubated at 37 °C for 24 h. The growth was calculated by spectrophotometer at 600 nm. MIC for bacteria was determined as the lowest concentration of the compound inhibiting the visual growth of the test cultures on the agar plate. Each test in this experiment was repeated three times and the average results in term of (mm of zone of inhibition) was tabulated.

3.4.4. Scanning Electron Microscopy

Scanning electron microscopy (SEM) was used to evaluate the susceptible bacteria that is subjected to the synthetic products. A Small piece of agar is cut from the inhibition zone and fixed in a glutaraldehyde solution 3% (*v/v*) buffered with 0.1 M sodium phosphate (pH 7.2) for 1 h at 25 °C. Later on, the samples are washed with a buffer of sodium phosphate. Then, all pieces were then post-fixed in 1 % (*w/v*) osmium tetroxide (OsO_4) for 1 h and then washed again in buffer. Dehydration steps were performed using propylene oxide. The specimens were left for drying and were mounted onto stubs using double-sided carbon tape, and then were coated with a thin layer of gold by a Polaron SC 502 sputter coater. All of them were observed on a JSM 6060 LV Scanning Electron Microscope (Jeol Inc., Peabody, MA, USA) [24].

4. CONCLUSIONS

A series of novel 2-amino-5-arylazothiazole disperse dyes was synthesized. The newly synthesized dyes were applied as disperse dyes for dyeing polyester fabric, where they exhibited good dye ability and fastness properties. The novel compounds displayed broad-spectrum antimicrobial efficiency against selected pathogenic Gram-positive and Gram-negative bacteria, as well as fungi. The remarkable activity against Gram-negative bacteria of the newly synthesized dyes will be subjected to further study for the action mechanism of each compound.

The antimicrobial activity of the synthesized dyes does not affect the dyeing properties when applied on fabrics but it is highly interesting to dye fabrics with effective antibacterial agents. Processes utilizing textile engineering and chemical technologies to improve fabrics that function on human skin have evolved over the centuries. Similar approaches are needed in fiber science to improve the quality and performance of health care fabrics.

ACKNOWLEDGMENTS

This work was supported by King Saud University, Deanship of Scientific Research, College of Science, Research Center.

AUTHOR CONTRIBUTIONS

H.E.G., M.M.G.F. and M.E.K. conceived, designed the experiments, performed the experiments, analyzed the data, contributed reagents/materials/analysis tools and wrote the paper.

REFERENCES

1. Dawson, J.F. Developments in disperse dyes. *Color. Technol.* **1978**, *9*, 25–35.

2. Malik, G.M.; Zadafiya, S.K. Thaizole based disperse dyes and their dyeing application on polyester fiber and their antimicrobial activity. *Chem. Sin.* **2010**, *1*, 15–21.

3. Ibatullin, U.G.; Petrushina, T.F.; Leitis, L.Y.; Minibaev, I.Z.; Logvin, B.O. Synthesis and transformations of 4-substituted 2-aminothiazoles. *Chem. Heterocycl. Compd.* **1993**, *29*, 612–615.

4. Gaffer, H.E.; Abdel-Latif, E. Antimicrobial activity of some new 4-arylazo-3-methylthiophene disperse dyes on polyester fabrics. *J. Appl. Polym.* **2011**, *122*, 83–89.

5. Beuchet, P.; Varache-Lembège, M.; Neveu, A.; Léger, J.M.; Vercauteren, J.; Larrouture, S.; Deffieux, G.; Nuhrich, A. New 2-sulfonamidothiazoles substituted at C-4: Synthesis of polyoxygenated aryl derivatives and *in*

vitro evaluation of antifungal activity. *Eur. J. Med. Chem.* **1999**, *34*, 773–779.

6. Geronikaki, A.; Vicini, P.; Dabarakis, N.; Lagunin, A.; Poroikov, V.; Dearden, J.; Modarresi, H.; Hewitt, M.; Theophilidis, G. Evaluation of the local anaesthetic activity of 3-aminobenzo[d]isothiazole derivatives using the rat sciatic nerve model. *Eur. J. Med. Chem.***2009**, *44*, 473–481.

7. Papadopoulou, C.; Geronikaki, A.; Hadjipavlou-Litina, D. Synthesis and biological evaluation of new thiazolyl/benzothiazolyl-amides, derivatives of 4-phenyl-piperazine.*Farmaco* **2005**, *60*, 969–973.

8. Gouda, M.; Gaffer, H.E.; Gouda, M.A. Synthesis and anti-hypertensive activity of novel sulphadimidine derivatives. *J. Med. Chem. Res.* **2012**, *21*, 3902–3906.

9. Kreutzberger, A.; Schimmelpfennig, H. Antivirale Wirkstoffe, 18. Mitt. 2-Aminothiazole durch Spaltung der S-S-Bindung des Disulfidodicarbamidins. *Arch. Pharm.* **1981**, *314*, 385–391.

10. Keil, D.; Flaig, R.; Schroeder, A.; Hartmann, H. Synthesis and characterisation of methine dyes derived from *N,N*-disubstituted 2-aminoselenazoles and some of their heterocyclic sulfur analogues. *Dyes Pigment.* **2001**, *50*, 67–76.

11. Metwally, M.A.; Abdel-latif, E.; Khalil, A.M.; Amer, F.A.; Kaupp, G. New azodisperse dyes with thiazole ring for dyeing polyester fabrics. *Dyes Pigment.* **2004**, *62*, 181–195.

12. Singh, K.; Singh, S.; Taylor, J.A. Monoazo disperse dyes—Part 1: Synthesis, spectroscopic studies and technical evaluation of monoazo disperse dyes derived from 2-aminothiazoles.*Dyes Pigment.* **2002**, *54*, 189–200.

13. Yen, M.S.; Wang, I.J. Synthesis and solvent characteristics of bishetaryl monoazo dyes derived from polysubstituted-2-aminothiophene derivatives. *Dyes Pigment.* **2005**, *67*, 183–188.

14. Khalifa, M.E.; Abdel-Latif, E.; Gobouri, A.A. Disperse dyes based on 5-arylazothiazol-2-ylcarbamoyl-thiophenes: Synthesis, antimicrobial activity and their application on polyester.*J. Heterocycl. Chem.* **2015**, *52*, 674–680.

15. Abdel-Wahab, B.F.; Gaffer, H.E.; Fahmy, H.M.; Fouda, M.E. Synthesis of some new 2-[(2,3-dihydroinden-1-ylidene) hydrazinyl]-4-methylthiazole derivatives for simultaneous dyeing and finishing for UV protective cotton fabrics. *J. Appl. Polym. Sci.* **2009**, *112*, 2221–2228.

16. Abdel-Latif, E.; Amer, F.A.; Metwally, M.A.; Khalifa, M.E. Synthesis of some 5-arylazo-2-(arylidenehydrazino)-thiazole disperse dyes for dyeing polyester fibers. *Pigment Resin Technol.* **2009**, *38*, 105–110.

17. Society of Dyes and Colorists. *Standard Methods for the Determination of the Color Fastness of Textiles and Leather*, 5th ed.; Society of Dyes and Colorists Publication: Bradford, UK, 1990.

18. Al-Etaibi, A.M.; El-Apasery, M.A.; Ibrahim, M.R.; Al-Awadi, N.A. A facile synthesis of new monoazo disperse dyes derived from 4-hydroxyphenylazopyrazole-5-amines: Evaluation of microwave assisted dyeing behavior. *Molecules* **2012**, *17*, 13891–13909.

19. Chipalkatti, H.R.; Desai, N.F.; Giles, C.H.; Macaulay, N. The influence of the substrate upon the light fading of azo dyes. *J. Soc. Dyers Color.* **1954**, *70*, 487–501.

20. Műller, C. Recent developments in the chemistry of disperse dyes and their intermediate.*Am. Dyest. Rep.* **1970**, *59*, 37–44.

21. Mcfarland, J. Nephelometer: An instrument for estimating the number of bacteria in suspensions used for calculating the opsonic index and for vaccines. *JAMA* **1907**, *14*, 1176–1178.

22. Chandrasekaran, M.; Venkatesalu, V. Antibacterial and antifungal activity of *Syzygium jambolanum* seeds. *J. Ethnopharmacol.* **2004**, *91*, 105–108.

23. Adesokan, A.; Akanji, M.A.; Yakubu, M.T. Antibacterial potentials of aqueous extract of*Enantia chlorantha* stem bark. *A. J. Biotechnol.* **2007**, *6*, 2502–2505.

24. Hayat, M.A. *Principles and Techniques of Electron Microscopy: Biological Applications*; University Park Press: Baltimore, MD, USA, 1981; Volume 1, p. 522.

Chapter 12

Effects of reductive stripping of reactive dyes on the quality of cotton fabric

Mohammad Gias Uddi[1*]n , Md. Mazedul Islam[2], Md. Rashedul Islam[1]

[1]Department of Textile Engineering, Ahsanullah University of Science and Technology, Tejgaon I/A, Dhaka 1208,Bangladesh.

[2]Department of Textile Engineering, Dhaka University of Engineering and Technology, Gazipur 1700, Bangladesh.

ABSTRACT

Some common problems of textile dyeing industries include uneven or faulty dyeing and formation of color patches on the fabric surface during dyeing and downstream processing of textiles materials. Such problems in the finished quality of fabric are generally tackled through a chemical stripping process which is a common practice in dyeing industries for the deep shade batches. However, reactive dyes cannot be stripped satisfactorily from cellulosic materials due to the formation of co-valent bonds between dye and fiber. This research was undertaken using 2.5% and 5% bi-hetero reactive dyes on pretreated cotton fabric and dye stripping was carried out in alkali reductive stripping process. The aim of the work was to investigate the effects of dye stripping on the quality of cotton fabric. Strength loss, weight loss, pilling resistance and absorbency of stripped fabric were calculated. Though with the increase of concentration of stripping chemicals and temperature, stripping percentages were improved; processing damage to the fabric such as losses in strength, weight and pilling resistance ratings was found. In contrast, increased fabric absorbency was found due to stripping. This is explained that during stripping, alkaline solution as an intracrystalline swelling agent is effective in loosening the crystalline region of cotton in addition to the amorphous region. Stripping agent can also attack such crystalline region. As a result, cotton fiber can release maximum number of hydroxyl groups which previously formed covalent bonds. This is the reason behind the stripped fabric having more water absorbency.

KEYWORDS

Reductive stripping, Fabric quality, Strength loss, Weight loss, Pilling, Fabric absorbency

1. INTRODUCTION

Around 10,000,000 tons of synthetic dyes are annually used in the world (Ghorpade et al. 2000; Goodarzian and Ekrami 2010; Saravanan et al. 2014) by various industries for coloring numerous materials. Textile industry accounts for two-thirds of the total dyestuff market (Anouzla et al. 2009; Riu et al. 1998; Shertate and Thorat2013). Over 10,000 dyes with an annual production of over 7×10^5 metric tons are commercially available for use by this industry (Campos et al. 2001). However, some common problems of textile dyeing industries include uneven or faulty dyeing and formation of color patches on the fabric surface during dyeing and downstream processing of textiles materials (Fono and Montclair 1980; Ogulata and Balci 2007). Such problems in the finished quality of fabric are generally tackled through a chemical stripping process which is a common practice in dyeing industries for the deep shade batches.

Stripping is one of the reproduction processes, used in textile finishing, removing dyes from colored fabric. The process is either termed as 'back stripping' or 'destructive stripping'. In back stripping only the depth of shade is changed while in destructive stripping the dyes are chemically altered (Ali et al. 2012; Fono and Montclair 1980), such as dyes containing an azo group (−N=N-) can be chemically reduced to an almost colorless amine compound by using chemical reducing agents. Sometimes for the analysis of dyes or dyed fabrics, dyes may need to be stripped from the surface by either destructive or non-destructive methods. Non-destructive stripping is essential when the amount of dye on the fiber has to be estimated, or the dye has to be identified or analyzed by techniques such as chromatography. Laboratory methods for stripping dyes seldom mirror those used in bulk processing, since these are usually destructive methods based on strongly reductive or oxidative methods (Park and Shore 2004). Depending on the individual dyes, complete chemical stripping may be carried out in one of the four ways: reduction only; reduction followed by oxidation; oxidation only; and oxidation followed by reduction (Aspland 1997). As sodium chlorite or peroxide solutions are used in oxidizing process, decreasing in strength of textile material is highly possible, so alkali reductive application is generally preferred for the stripping of reactive dyes. The mechanism of reductive stripping depends on the type of dyes, fibers and reducing agents. Dyes are commonly stripped from fabric using a reducing agent, and a stripping assistant which is used to enhance the stripping efficiency of a reducing agent. Various chemical combinations of reducing agents and stripping assistants are being used in stripping liquids to strip the dye from fabric (Chavan 1969).

The reductive stripping occurs from chemical reactions in case of cellulose dyed with reactive dyes. The chemical bonds on the chromophore group of reactive dyes are tried to break in order to dissolve of colorfulness. Reactive dyes cannot be stripped satisfactorily from cellulosic materials, due to the formation of co-valent bond between dye and fiber (Choudhury 2006). The stripping performance depends on several factors: raw materials, type of dyestuff, method of dyeing and stripping, type of stripping agents and auxiliaries, and working parameters such as time, temperature etc. Achieving the expected efficiency in the application of stripping process is a very important aspect for the success of reproduction (Ogulata and Balci 2007). However, alkaline reductive stripping also affects the quality of fabric as harsh chemicals such as sodium hydrosulphite and sodium hydroxide as well as high temperature is applied in the stripping process, but the effects are at what extent? The aim of this study was to evaluate such effects of dye stripping on the quality of fabric.

2. METHODS

2.1 Materials

Pretreated (Scoured-bleached) cotton fabric (Single jersey) of about 166 g/m^2 (Yarn count 25^S Ne), supplied by Hi-Fashion Composite Textiles Ltd., Joydevpur, Gazipur was used throughout the study. Bi-hetero type (Flurotriazine and vinyl sulphone) reactive dyes: Novacron Red TS-3B, Novacron Yellow TS-3R and Novacron Blue TS-3G supplied by Huntsman Bangladesh Ltd. was used for this research purpose. Sodium hydrosulphite supplied by Titas Spinning and Denim Company Ltd. Mawna-Bhabanipur, Gazipur was used as a stripping agent. All other chemicals and auxiliaries used for the experimental work were laboratory grade reagents.

2.2 Dyeing process

Dyeing was carried out on the pretreated fabric with the two shades (2.5% and 5%) using combination shade in exhaust method in the IR sample dyeing machine at 60°C for 40 min, keeping material to liquor ratio 1:10 and pH 10.5. Then after washing, the dyed fabric was neutralized with 1 g/l acetic acid followed by soaping with 1 g/l Dekol SN at 80°C for 5 min. Dyeing recipes for different shade% of reactive dyes was given in Table 1.

Table 1. Dyeing recipes using 2.5% and 5% reactive dyes

Sl. no	Shades			Recipe			
	Dyes	Combined shade	Individual shade	Glauber's salt (g/l)	Soda ash (g/l)	Wetting agent (g/l)	Sequestering agent (g/l)
1.	Novacron red TS-3B	2.5%	2.0%	60	14	1	1
	Novacron yellow TS-3R		0.4%				
	Novacron blue TS-3G		0.1%				
2.	Novacron red TS-3B	5%	4.0%	100	18		
	Novacron yellow TS-3R		0.8%				
	Novacron blue TS-3G		0.2%				

2.3 Stripping process

Eight different stripping operations (see Table 2) were carried out for each shade% in the sample dyeing machine using material to liquor ratio 1:15 for 30 min.

Table 2. Stripping recipes

Stripping operations	Temperature (°C)	Conc. of sodium hydrosulphite (g/l)	Conc. of caustic soda (g/l)
1.	80	5	2.5
2.		5	5
3.		10	5
4.		10	10
5.	100	5	2.5
6.		5	5
7.		10	5
8.		10	10

2.4 Stripping percentage

The color yields of both dyed and stripped fabric were determined in terms of K/S value by a dual beam reflectance spectrophotometer. The stripping percentage was calculated using equation (1) (Ali et al. 2012).

$$\text{Stripping percentage} = \frac{[\text{K/S value of dyed sample} - \text{K/S value of stripped sample}}{\text{K/S value of dyed sample}} \times 100 \quad (1)$$

2.5 Fabric quality evaluation

2.5.1 Strength loss

Bursting strength of dyed and stripped fabric was carried out in ISO 13938–1:1999 method by Bursting Strength Tester. Then strength loss of the stripped sample was measured using equation (2).

$$\text{Strength loss percentage} = \frac{(\text{Fabric strength before stripping} - \text{Fabric strength after stripping})}{(\text{Fabric strength before stripping})} \times 100$$

(2)

2.5.2 Weight loss

Loss in weight of stripped sample was measured by difference in weight before and after treatment using equation (3).

$$\text{Weight loss percentage} = \frac{(\text{Fabric weight before stripping} - \text{Fabric weight after stripping})}{(\text{Fabric weight before stripping})} \times 100$$

(3)

2.5.3 Pilling

Pilling resistance of stripped fabric samples was assessed in SN 198525 method using Martindale Abrasion and Pilling Tester.

2.6 Fabric water absorbency

Vertical wicking test of fabric was carried out using "Ciba" Test Method reported by Kate Spade & Company (2014). The 15 cm × 2.5 cm strips of samples (Both in wale and coarse direction) was lowered to touch the water reservoir at which moment timer was started, and the position of water front line (cm) over 5 min was recorded.

3. RESULTS AND DISCUSSION

3.1 Stripping percentage

Stripping percentages of the dyed fabrics for 2.5% and 5% shades were displayed in Figures 1 and 2, respectively.

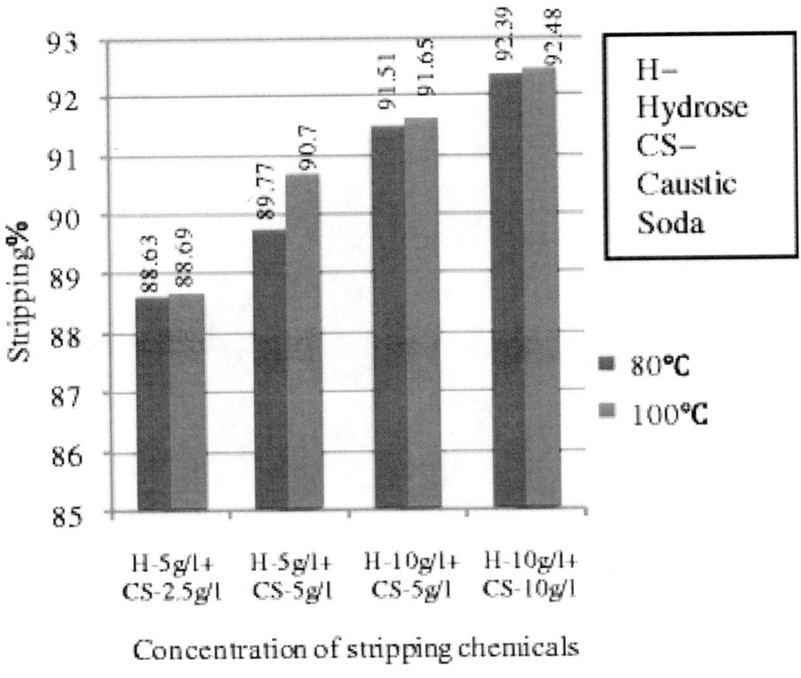

Figure 1. Stripping percentages of dyed fabric (2.5% shade).

From Figure 1, for 2.5% shade, it was found that stripping percentages of dyed fabrics increased gradually with the increase of concentration of stripping chemicals (Hydrose and caustic soda) when fabrics were stripped both at 80°C and 100°C. Stripping percentages were also found higher at 100°C than 80°C under each combination of stripping chemicals. As the temperature of the stripping operation was more, the stripping was more. Both at 80°C and 100°C temperature when concentration of caustic soda was doubled from 2.5 g/l to 5 g/l with 5 g/l hydrose as well as from 5 g/l to 10 g/l with 10 g/l hydrose, stripping percentages were slightly increased.

From Figure 2, similar trends were found for 5% shade. Using 10 g/l hydrose and 10 g/l caustic soda and stripping at 100°C, the maximum stripping for 2.5% and 5% shades were found 92.8422% and 93.801%, respectively.

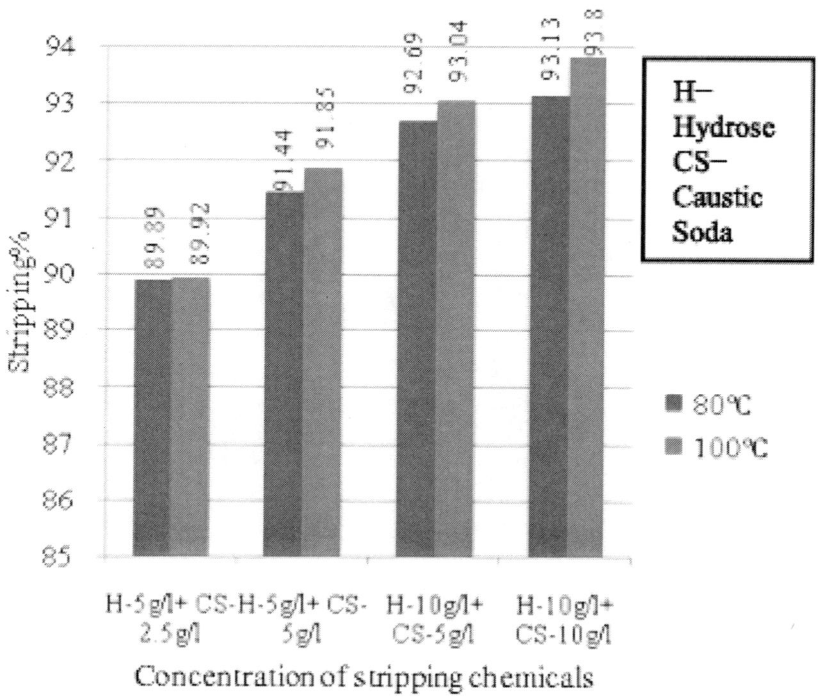

Figure 2. Stripping percentages of dyed fabric (5% shade).

3.2 Strength loss

Strength loss percentages of the dyed fabrics for 2.5% and 5% shades were shown in Figures 3 and 4, respectively.

In case of 2.5% shade, strength loss percentages were increased gradually with the increase of concentration of stripping chemicals both at 80°C and 100°C as shown in Figure 3. In between 80°C and 100°C, it was noticed that strength loss percentages were higher at 100°C when the fabrics were treated in same parameter. Here, the sample dyed with 2.5% shade when treated with 10 g/l hydrose and 10 g/l caustic soda at 100°C showed maximum strength loss of 9.23% in comparison to the maximum strength loss of 8.46% when treated at 80°C under the same combination of stripping chemicals. As both the temperature and stripping chemicals were comparatively more, the extent of strength loss was also found more. Again, when caustic soda was doubled from 2.5 g/l to 5 g/l with 5 g/l hydrose, strength losses were increased from 3.07% to 4.61% (In case of 80°C) and from 3.84 to 5.38% (In case of 100°C). Similarly, increasing caustic soda from 5 g/l to 10 g/l with 10 g/l hydrose resulted in increased strength loss from 6.15% to 8.46% (In case of 80°C) and from 7.69% to 9.23% (In case of 100°C).

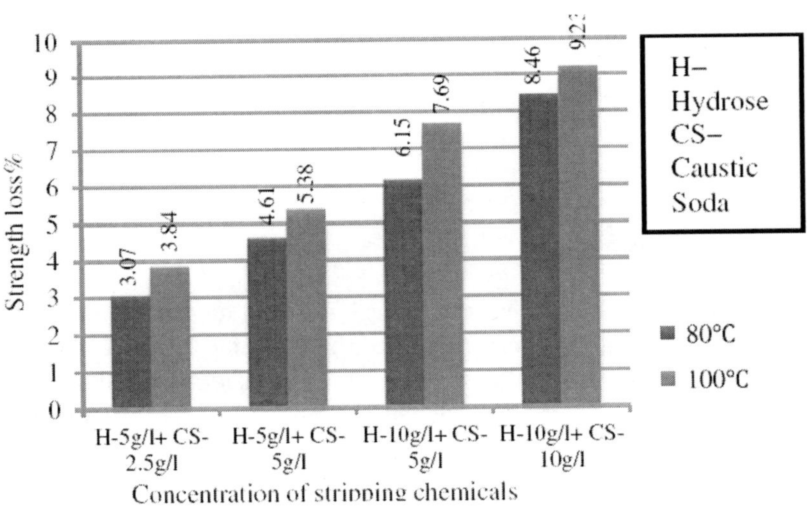

Figure 3. Strength loss% of stripped samples (2.5% shade).

For 5% shade, similar trends of gradual increasing of strength loss with the increase of concentrations of stripping chemicals were found as shown in Figure 4 for both 80°C and 100°C. The samples dyed with 5% shade when stripped using 10 g/l hydrose and 10 g/l caustic soda at 100°C showed maximum strength loss of 10.0% in comparison to the maximum strength loss of 9.23% when treated at 80°C under the same combination of stripping chemicals.

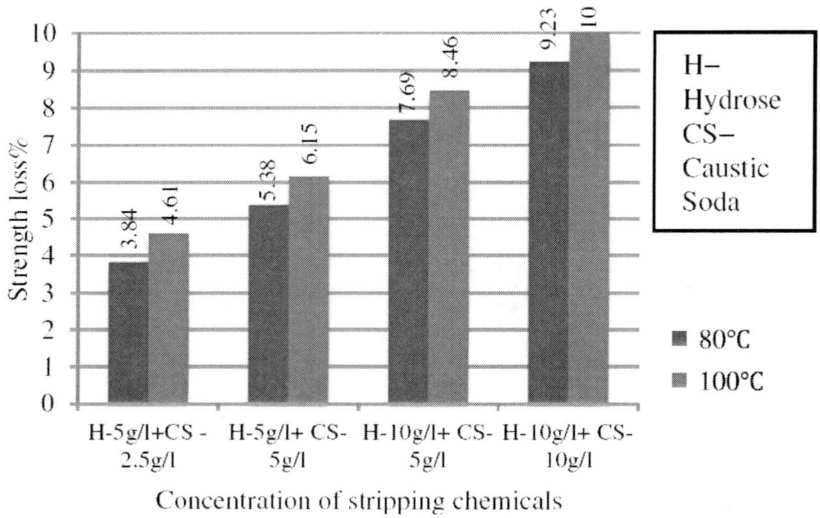

Figure 4 Strength loss% of stripped samples (5% shade).

3.3 Weight loss

Quality of stripped fabric was also evaluated by measuring the percent weight loss. Weight loss percentages of the dyed fabrics for 2.5% and 5% shades were displayed in Figures 5 and 6, respectively. From Figure 5, for 2.5% shade it was noticed that weight loss percentages of stripped fabrics increased with the increase of concentration of stripping chemicals when fabrics were stripped both at 80°C and 100°C. The weight loss percentages were found higher at 100°C than 80°C under each combination of stripping chemicals used. As the temperature of stripping operations was more, the weight loss percentage was found more. Here, the sample dyed with 2.5% shade when treated with 10 g/l hydrose and 10 g/l caustic soda at 100°C showed maximum weight loss of 4.91% in comparison to the maximum weight loss of 4.8% when treated at 80°C under the same combination of stripping chemicals.

In case of 5% shade, similar trend was found as shown in Figure 6. The samples dyed with 5% shade when stripped using 10 g/l hydrose and 10 g/l caustic soda at 100°C showed maximum weight loss of 5.99% in comparison to the maximum weight loss of 5.76% when treated at 80°C under the same combination of stripping chemicals.

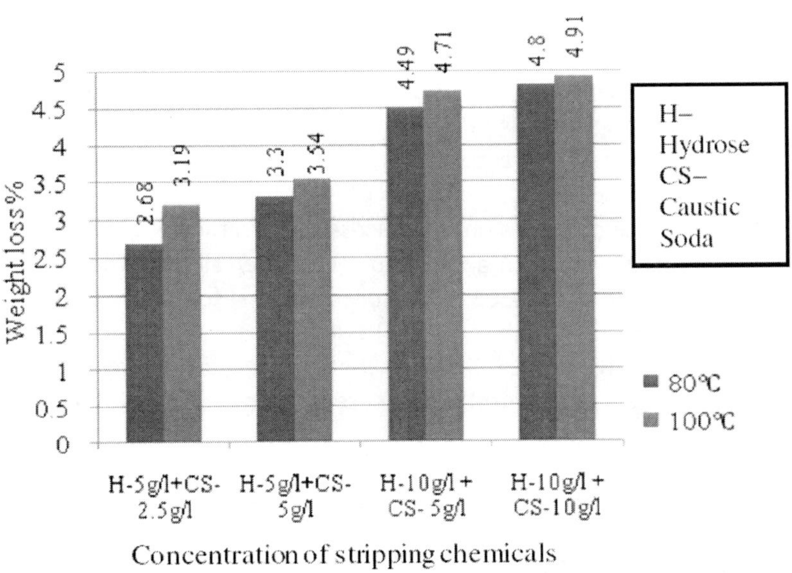

Figure 5. Weight loss% of stripped fabric (2.5% shade).

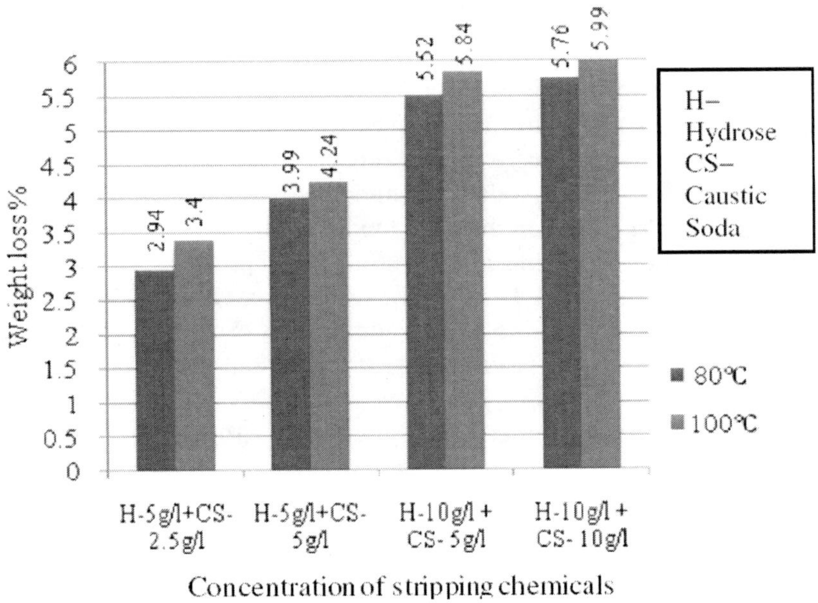

Figure 6. Weight loss% of stripped fabric (5% shade).

Reactive dyes are the most substantive of dyes used on cotton because of their covalent bonding with glucoside hydroxyls. As a result, these dyes are the most resistant to extraction or destruction of all dye types. An aqueous solution of a strong base is typically used to remove the dyes from the cotton fiber. Complete extraction of reactive dyes was reported from bulk cotton treated with 1.5% aqueous NaOH at 100°C for 20 min (Dockery et al. 2009). The alcohol groups on the glucose units in the cellulose backbone of cotton act as a weak acid and are ionized under alkaline conditions. As cotton fibers partially dissolve in concentrated sodium hydroxide at elevated temperatures, fabric strength loss occurred. As such, during alkaline dye reductive process fabric strength loss occurred. Strength loss of stripped fabric increased with the increase of stripping action provided by the elevated temperature and concentration of hydrose and caustic soda. Increased weight losses of stripped fabrics are the consequences of higher strength loss due to similar way.

3.4 Pilling resistance

3.4.1 Ratings for dyed and stripped samples in case of 2.5% shade

Pilling resistance ratings were found 5, 4-5 and 4 under 125, 500 and 2000 cycles respectively for fabric dyed with 2.5% dyes as shown in Figures 7 and 8. Ratings were found similar as dyed fabric when 5 g/l hydrose and 2.5 g/l caustic soda combination were used at 80°C but ratings decreased by 1/2 grade than the dyed fabric for 500 and 2000 cycles under same combination of chemicals at 100°C. In other three combinations, ratings were found 4-5, 4, and 3-4 when treated at 125, 500 and 2000 cycles respectively for both the temperatures except the combination of 10 g/l hydrose and 10 g/l caustic soda at 100°C (Rating 3).

For both the stripping temperatures, in case of 5 g/l hydrose and 2.5 g/l caustic soda combination, pilling resistance ratings were 5, 4-5 and 4 under 125, 500 and 2000 cycles respectively whereas for 5 g/l hydrose and 5 g/l caustic soda, corresponding ratings were 5, 4-5 and 3-4. When the dyed sample was treated with 10 g/l hydrose and 5 g/l caustic soda or 10 g/l hydrose and 10 g/l caustic sodacombination at both 80°C and 100°C, pilling ratings were found 4-5, 4 and 3-4 under 125, 500 and 2000 cycles respectively, except rating 3 was found under 2000 cycles for 10 g/l hydrose and 10 g/l caustic soda combination stripping treated at 100°C. The alkali-treated cotton surfaces appeared fuzzy and blurred at the time of reductive stripping at high temperature and so lower pilling resistance.

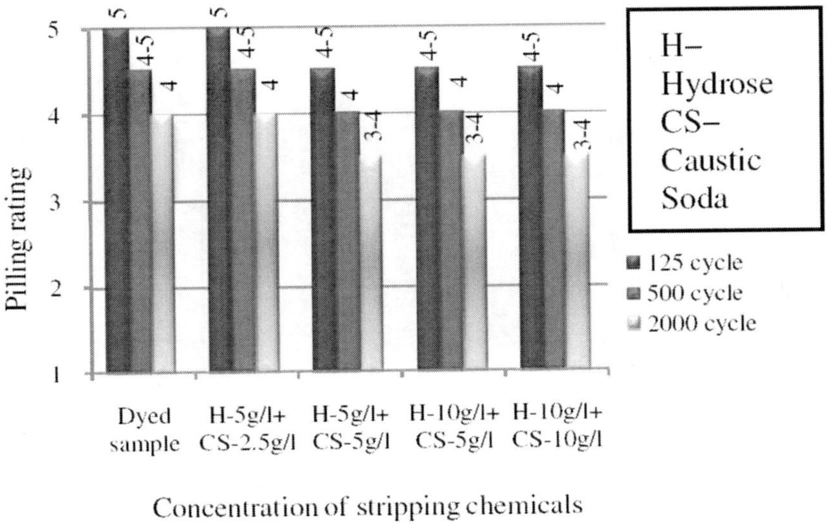

Figure 7. Pilling ratings of stripped samples at 80°C (2.5% shade).

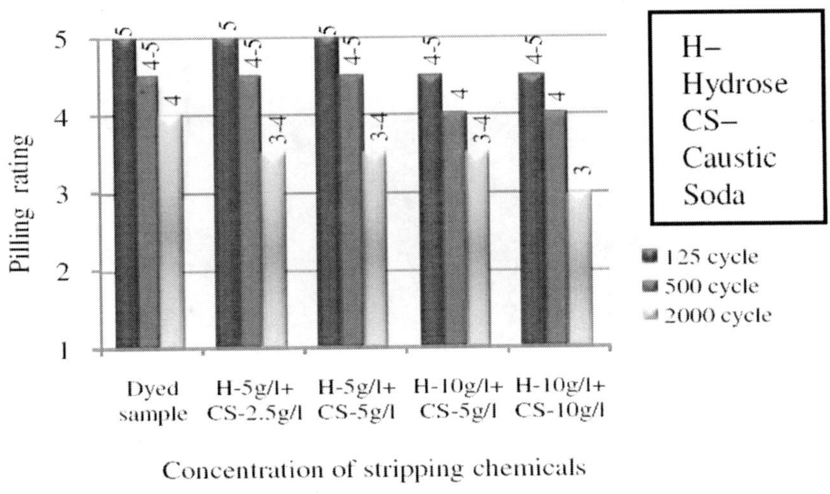

Figure 8. Pilling ratings of stripped samples at 100°C (2.5% shade).

3.4.2 Ratings for dyed and stripped samples in case of 5% shade

For 5% shade, pilling resistance ratings of dyed fabric were found 5 under 125 cycles and 4-5 under 500 and 2000 cycles as shown in Figures 9 and 10.

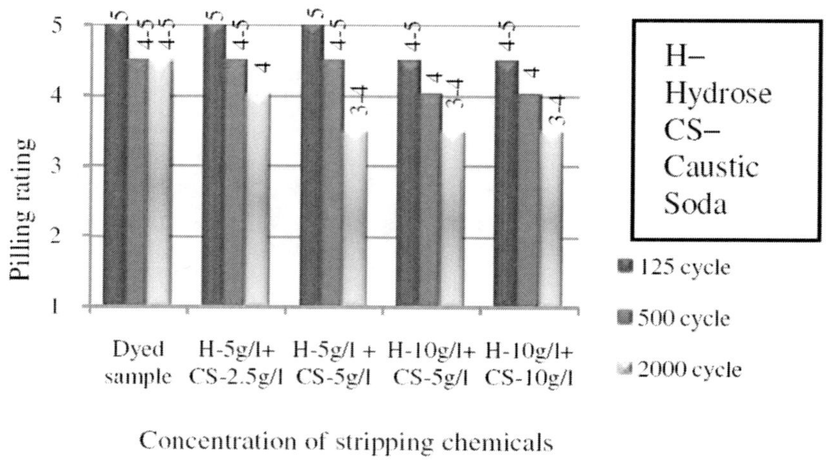

Figure 9. Pilling ratings of stripped samples at 80°C (5% shade).

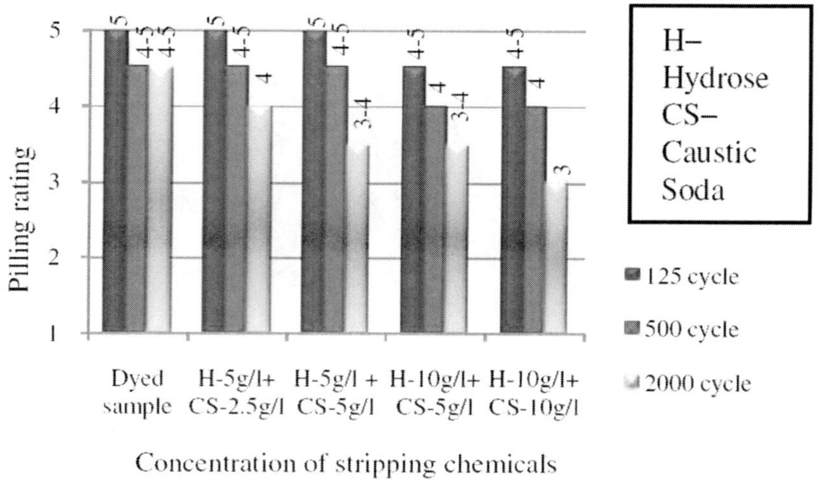

Figure 10. Pilling ratings of stripped samples at 100°C (5% shade).

3.5 Fabric absorbency

The wicking test results of pretreated, dyed and stripped fabrics were shown in Table 3. The average absorbing length was found 9.5 cm for pretreated fabric in both wale and coarse direction, whereas it increased to 10.5 cm (In wale direction) and 10 cm (In coarse direction) for dyed fabrics and further increased in the stripped fabrics. For stripped fabrics, it ranged within 11.5 cm to 14 cm in wale direction and 11 cm to 13.7 cm in coarse direction for 2.5% shade. Again, for 5% shade, it ranged within 11.5 cm to 13.9 cm in wale direction and 10.6 cm to 13.9 cm in coarse direction.

Table 3. Wicking test results of pretreated, dyed and stripped fabric

Shade%	Stripping recipes			Water front line (cm)					
	Temperature (°C)	Hydrose (g/l)	Caustic Soda (g/l)	Pretreated fabric		Dyed fabric		Stripped fabric	
				Wale	Course	Wale	Course	Wale	Course
2.5	80	5	2.5	9.5	9.5	10.5	10	13.5	12.5
		5	5					12.25	12.25
		10	5					11.5	11
		10	10					12	11.5
	100	5	2.5					13.9	12.9
		5	5					14	13.7
		10	5					12.7	12.1
		10	10					13.1	12.5
5.0	80	5	2.5					13	12.25
		5	5					13.5	13
		10	5					13.9	13.5
		10	10					13.4	13
	100	5	2.5					12.2	10.65
		5	5					13	11.6
		10	5					11.5	11.5
		10	10					12	10.6

Thus it can be said from the wicking test results that the water front line in the stripped samples was pronouncedly higher than the pretreated one. The better wettability of the stripped samples might be due to the stripping treatment under alkaline condition. In addition, the greater absorbency of the dyed fabric in comparison to the pretreated fabric was due to dyeing of cotton fabric with reactive dyes was carried out in alkaline media (Liu and Wang 2009). Both dyed and stripped fabrics were treated under different concentrations of alkali, material to liquor ratio, temperature and time; but in case of stripping process of the dyed fabrics higher concentration of alkali and temperature were used than the dyed ones and correspondingly found higher wicking distance in the stripped fabrics.

Moreover, water is an intercrystalline swelling agent which can penetrate and loosen only the amorphous region of cellulose; while alkaline solution, as an intracrystalline swelling agent affects both the amorphous and crystalline regions of cellulose. As a result, alkaline solution is effective in loosening the crystalline region of cellulose. The action of such swelling agent inherits the outer skin on cotton fibers and causes it to split and form collars; the inner cellulose layers swell rapidly the collars (Fan et al. 1987). Due to loosening of crystalline region of cellulose by swelling agents, the absorbency of the fabric towards water and dyes is increased.

It can be explained in other words. Cotton is pure cellulose, a naturally occurring polymer. Chemically it is a "cellobiose" which belongs to carbohydrate polymer, containing hydroxyl groups. These negatively charged groups attract water molecules and make cotton absorbs water well. When cotton dyed with reactive dyes, some of these -OH groups are occupied, which slightly reduces absorbency. During the stripping process, cotton fiber releases the maximum number of hydroxyl groups which has been formed covalent bonds with the cellulose. This is due to the dye reduction reaction of sodium hydrosulphite and caustic soda on the cellulose. In addition, the stripping agent can also attack the crystalline region of cotton and as a result, there may also be chances of releasing the hydroxyl groups. Basically, hydroxyl groups are responsible for water absorbency. This is the reason behind the stripped fabric having more water absorbency as well as strength loss.

Furthermore, the basic preparation processes of cotton knit fabric include scouring and bleaching. These treatments remove natural and human induced impurities, that is, noncellulosic constituents and other unwanted substances. Thus increase the affinity of cellulose for dyes and finishes. It may also be possible that any impurities remained in the fabric could not be removed at the elementary preparation processes and they also remained in the dyed fabrics, which were finally removed due to alkaline reductive stripping action. As a result, absorbency of the stripped fabric might also be increased showing higher wicking distant.

4. CONCLUSIONS

In this study, an investigation was carried out on alkali reductive stripping process for the cotton knit fabric dyed with the Novacron Red TS-3B, Novacron Yellow TS-3R and Novacron Blue TS-3G reactive dyes concerning the working parameters as the concentration of reducing agent and caustic soda, stripping temperature. Based on this experimental study, the conclusions drawn were:

1 Fabric strength loss percentages increased gradually with the increase of concentration of stripping chemicals and temperature of stripping bath. The sample dyed with 2.5% and 5% shades when treated with 10 g/l hydrose and 10 g/l caustic soda at 100°C showed maximum strength loss of 9.23% and 10.0% respectively.

2 Fabric weight loss percentages of stripped fabrics were increased gradually with the increase of concentration of stripping chemicals and stripping temperature. The sample dyed with 2.5% and 5% shades when treated with 10 g/l hydrose and 10 g/l caustic soda at 100°C showed maximum weight loss of 4.91% and 5.99%, respectively.

3 In most cases, stripped samples showed lower pilling resistance rating than the corresponding dyed fabric.

4 Fabric absorbency was increased considerably due to stripping.

The stripping of reactive dyes from cotton knit fabric and its satisfactory performance level directly related to the processing damage to the fabric. Besides the better striping percentage, the overall fabric quality needs to be considered at the time of stripping. As stripping is a dye destruction process using harsh chemicals, there is no way to stop completely their adverse effects on fabric quality, but this can be controlled within a certain range. In case of major unevenness 5 g/l hydrose and 5 g/l alkali is commonly used in knit dyeing industries for full stripping of reactive dyes from cotton fabric and according to this study the fabric strength loss and weight loss can be controlled within 5% for both using such combinations.

Stripping whether partial or full should be selected at the elementary stage in the bulk production on the basis of major dyeing faults and expected target shade. In general, dyeing industries take decision of dye stripping when it becomes absolutely necessary for their not ok dyed fabrics. For such not ok, already rejected to delivery batches, manufacturers get ready to compromise with the quality of fabric up to a controlled limit. If stripping can be carried out under optimum alkaline condition which is just sufficient to maintain the stability of sodium hydrosulphite in the stripping bath will definitely reduce the fabric damaging effects. Unevenness can be removed from the surface of the fabric through stripping and then re-dyeing can give the fabric new look as expected shade, as if it can be said that stripping is an aspect of improving the appearance and evenness.

ACKNOWLEDGEMENT

The authors of the paper would like to thank Mr. Rezaul Hasan (Rumman), Manager, Concept Knitting Ltd., Masco Group for his technical support.

AUTHORS' CONTRIBUTIONS

MGU designed and carried out the experiments, analyzed the data as well as drafted the manuscript. All authors read and approved the final manuscript.

REFERENCES

1. Ali, S, Chathaa, S, Asghera, M, Ali, SK, & Hussain, AI. (2012). Biological color stripping: a novel technology for removal of dye from cellulose fibers. *Carbohydr Polym, 87*(2), 1476–1481.

2. Anouzla, A, Souabi, S, Safi, M, Aboulhassan, A, Rhbal, H, & Abrouki, Y. (2009). Valorization of steel industry wastewaters in the decolorization of dyes containing solutions. *Scientific Study & Research, 10*(3), 277–284.

3. Aspland, JR. (1997). *Textile dyeing and coloration (p. 130)*. Research Triangle Park, NC: American Association of Textile Chemists and Colorists.

4. Campos, R, Kandelbauer, A, Robra, KH, Artur, CP, & Gubitz, GM. (2001). A rapid and sensitive method for the quantification of protein using the principle of protein-dye binding. *Anal Biochem, 72*, 248–254.

5. Chavan, RB (1969). *Stripping of dyestuff from textiles, US Patent No. 3591325.*

6. Choudhury, RKA. (2006). *Textile preparation and dyeing (p. 550)*. New Delhi, India: Oxford and IBH Publishing Ltd.

7. Dockery, CR, Stefan, AR, Nieuwland, AA, Roberson, SN, Baguley, BM, Hendrix, JE, & Morgan, SL. (2009). Automated extraction of direct, reactive, and vat dyes from cellulosic fibers for forensic analysis by capillary electrophoresis. *Anal Bioanal Chem, 394*, 2095–2103.

8. Fan, LT, Gharpurag, MM, & Lee, YH. (1987). *Biotechnology monograph: cellulose hydrolysis volume vi (p.14–17, 66)*. New York: Springer Verlag, Berlin.

9. Fono, A, & Montclair, NJ (1980 October 14). *New process of color stripping dyed textile fabrics. US Patent No. 4227881.*

10. Ghorpade, B, Darvekar, M, & Vankar, PS. (2000). Ecofriendly cotton dyeing with Sappan wood dye using ultrasound energy. *Colorage, 45*(1), 27–30.

11. Goodarzian, H, & Ekrami, E. (2010). Extraction of dye from madder plant (Rubia tictorium L) and dyeing of wool. *World Appl Sci J, 9*(4), 434–436.

12. Kate Spade & Company. (2014 February). *Standard test methods: wicking test methods,* p. 6. Retrieved December 29, 2014 from https://portal.katespadeconnect.com/documents/10113/1288847/Standard+Test+Methods

13. Liu, J, & Wang, F-m. (2009). Effect of alkalization treatment on structure and properties of bleached kapok yarn. *Journal of Xi'an Polytechnique University, 23*(2), 386–394.

14. Ogulata, RT, & Balci, O. (2007). Investigation of the stripping process of the reactive dyes using organic sulphur reducing agents in alkali condition. *Fibers Polym, 8*(1), 25–36.

15. Park, J, & Shore, J. (2004). *Practical dyeing volume ii*. England: The Society of Dyers and Colorists.

16. Riu, J, Schonsee, I, & Barcelo, D. (1998). Determination of sulfonated azo dyes in groundwater and industrial effluents by automated solid-phase extraction followed by capillary electrophoresis/mass spectrometry. *J Mass Spectro, 33*, 653–663.

17. Saravanan, P, Chandramohan, G, Mariajancyrani, J, & Kiruthikajothi, K. (2014). Eco-friendly dyeing of wool fabric with a natural dye extracted from barks of Odinawodier. *Der Chemica Sinica, 5*(1), 28–33.

18. Shertate, RS, & Thorat, PR. (2013). Biotransformation of sulphonated azo dye direct red 5B by marinobacter sp.dr-7- a bioremedial aspect in marine environment. *Int J Pharma Bio Sci, 4*(3B), 524–534.

Index